清華大学 计算机系列教材

沈美明 温冬婵 张赤红 编著

IBM-PC汇编语言程序设计
实验教程

清华大学出版社
北京

内 容 简 介

本书共分五章：第一章介绍上机的基本方法，特别强调了 DEBUG 的使用；第二章为基本程序结构（循环、分支和子程序）训练；第三章介绍几种主要输入/输出设备的编程技术；第四章说明磁盘文件存取技术；第五章为以宏为主的高级汇编语言技术及连接技术训练。全书共给出了 22 个例题及 17 个实验题。这组实验的综合性较强。它综合了顺序、循环、分支和子程序四种基本结构的编程技术，同时又增加了系统功能调用、BIOS 调用、宏汇编及条件汇编功能、模块连接技术及中断程序设计技术等内容。它又包括了表格查找、声音输出、键盘输入、显示及窗口技术、画线技术以及顺序式、随机式、文件代号式磁盘文件存取技术等汇编语言最经常使用的场合所需要的技术。因此，这组实验对学生的训练是全面的。而对于工程技术人员则可以各取所需，协助解决你所遇到的实际问题。

本书封面贴有清华大学出版社防伪标签，无标签者不得销售。
版权所有，侵权必究。侵权举报电话：010-62782989　13701121933

图书在版编目(CIP)数据

IBM-PC 汇编语言程序设计实验教程/沈美明，温冬婵，张赤红编著. —北京：清华大学出版社，1992(2020.1重印)
清华大学计算机系列教材
ISBN 978-7-302-01033-3

Ⅰ. ①I… Ⅱ. ①沈… ②温… ③张… Ⅲ. ①汇编语言－程序设计－高等学校－教材 Ⅳ. ①TP313

中国版本图书馆 CIP 数据核字(2007)第 057816 号

责任编辑：	贾仲良　白立军
封面设计：	刘艳芝
责任校对：	焦丽丽
责任印制：	刘海龙

出版发行：清华大学出版社
　　网　　址：http://www.tup.com.cn, http://www.wqbook.com
　　地　　址：北京清华大学学研大厦 A 座　　　邮　　编：100084
　　社 总 机：010-62770175　　　　　　　　　邮　　购：010-62786544
　　投稿与读者服务：010-62776969, c-service@tup.tsinghua.edu.cn
　　质量反馈：010-62772015, zhiliang@tup.tsinghua.edu.cn

印 装 者：北京国马印刷厂
经　　销：全国新华书店
开　　本：185mm×260mm　　印　张：12.75　　字　数：292 千字
印　　次：2020 年 1 月第 48 次印刷
定　　价：29.00 元

产品编号：001033-04

前　言

　　本书与清华大学出版社已出版的《IBM-PC 汇编语言程序设计》和《IBM-PC 汇编语言习题集》一起组成配套教材，主要面向高等院校"汇编语言程序设计"的课程教学。"汇编语言程序设计"是一门实践性很强的课程，只有通过上机实践才有可能掌握程序设计技术并使其达到较高的水平，因此我们编写了这本实验教程，一方面为"汇编"课程的实验教学服务；另一方面，需要使用汇编语言的工程技术人员也可以根据本书内容进行上机实践，从中得到收益及提高。

　　本书共分五章：第一章介绍上机的基本方法，特别强调了 DEBUG 的使用；第二章为基本程序结构（循环、分支和子程序）训练；第三章介绍几种主要输入/输出设备的编程技术；第四章说明磁盘文件存取技术；第五章为以宏为主的高级汇编语言技术及连接技术训练。全书共给出了 22 个例题及 17 个实验题。这组实验的综合性较强。它综合了顺序、循环、分支和子程序四种基本结构的编程技术，同时又增加了系统功能调用、BIOS 调用、宏汇编及条件汇编功能、模块连接技术及中断程序设计技术等内容。它又包括了表格查找、声音输出、键盘输入、显示及窗口技术、画线技术以及顺序式、随机式、文件代号式磁盘文件存取技术等汇编语言最经常使用的场合所需要的技术。因此，这组实验对学生的训练是全面的。而对于工程技术人员则可以各取所需，协助解决你所遇到的实际问题。

　　在清华大学计算机系的"汇编"课中，安排了 32 机时的上机训练。要求学生完成 10 个属于基本要求的实验题，其余 7 个实验题（带 * 号）供有余力的学生选做。各兄弟院校可以根据自己的机时安排选择部分实验题供学生实验用。

　　本书的第一、二章及附录由沈美明编写，第三章由温冬婵编写，第四、五章由张赤红编写。书中如有错误和不当之处，欢迎读者批评指正。

目 录

第一章　实验的基本要求与方法 ……………………………………………… 1
  1.1　实验目的与要求 ……………………………………………………… 1
    一、实验目的 …………………………………………………………… 1
    二、实验要求 …………………………………………………………… 1
  1.2　实验方法 ……………………………………………………………… 2
    例 1.1　比较字符串 sample …………………………………………… 2
第二章　程序的基本结构练习 ………………………………………………… 13
  2.1　循环程序设计 ………………………………………………………… 13
    一、示例 ………………………………………………………………… 13
      例 2.1　表格查找 tabsrch ………………………………………… 13
      例 2.2　建立学生名次表 rank …………………………………… 16
    二、实验题 ……………………………………………………………… 19
      实验 2.1　用表格形式显示字符 ………………………………… 19
      实验 2.2　查找匹配字符串 ……………………………………… 20
  2.2　分支程序设计 ………………………………………………………… 21
    一、示例 ………………………………………………………………… 21
      例 2.3　统计学生成绩 result …………………………………… 21
      例 2.4　显示月份名 direct ……………………………………… 24
      例 2.5　显示错误信息 show_err ………………………………… 27
    二、实验题 ……………………………………………………………… 34
      实验 2.3*　分类统计字符个数 ………………………………… 34
  2.3　子程序设计 …………………………………………………………… 34
    一、示例 ………………………………………………………………… 34
      例 2.6　显示学生名次表 rank …………………………………… 34
      例 2.7　计算工资 scremp ………………………………………… 40
      例 2.8　HANOI 塔题 ……………………………………………… 52
    二、实验题 ……………………………………………………………… 67
      实验 2.4　查找电话号码 ………………………………………… 67
      实验 2.5*　求 Fibonacci 数 ……………………………………… 68
第三章　I/O 程序设计 ………………………………………………………… 70
  3.1　发声系统程序设计 …………………………………………………… 70
    一、示例 ………………………………………………………………… 71
      例 3.1　枪声程序 gun …………………………………………… 71

　　　　例 3.2　演奏音阶程序 musex ……………………………………… 73
　　二、实验题 ……………………………………………………………… 74
　　　　实验 3.1*　乐曲程序(1) ………………………………………… 74
　　　　实验 3.2　乐曲程序(2) …………………………………………… 76
3.2　显示器 I/O 程序设计 …………………………………………………… 76
　　一、示例 …………………………………………………………………… 78
　　　　例 3.3　光标轨迹程序 draw ……………………………………… 78
　　　　例 3.4　窗口控制程序 wdex ……………………………………… 81
　　　　例 3.5　画横竖线程序 grid ……………………………………… 86
　　二、实验题 ……………………………………………………………… 88
　　　　实验 3.3*　字符图形程序 ………………………………………… 88
　　　　实验 3.4　屏幕窗口程序 ………………………………………… 89
　　　　实验 3.5*　画栅栏线程序 ………………………………………… 91
3.3　键盘输入程序设计 ……………………………………………………… 91
　　一、示例 …………………………………………………………………… 93
　　　　例 3.6　键盘处理演示程序 kbdio ………………………………… 93
　　　　例 3.7　键盘输入程序 keyboard ………………………………… 97
　　　　例 3.8　字处理演示程序 wspp …………………………………… 100
　　二、实验题 ……………………………………………………………… 105
　　　　实验 3.6　扩充键盘处理功能的程序 …………………………… 105
　　　　实验 3.7*　扩充字处理功能的程序 ……………………………… 106
3.4　中断程序设计 …………………………………………………………… 107
　　一、示例 …………………………………………………………………… 108
　　　　例 3.9　打字计时程序 type_ex …………………………………… 108
　　二、实验题 ……………………………………………………………… 116
　　　　实验 3.8　中断练习程序 ………………………………………… 116

第四章　文件管理 …………………………………………………………… 118
4.1　文件代号方式下的文件管理 …………………………………………… 118
　　一、示例 …………………………………………………………………… 118
　　　　例 4.1　分页显示文件 ex_41 ……………………………………… 118
　　　　例 4.2　删除页 ex_42 ……………………………………………… 126
　　二、实验题 ……………………………………………………………… 137
　　　　实验 4.1　页拷贝 ………………………………………………… 137
4.2　文件控制块方式下的文件管理 ………………………………………… 137
　　一、示例 …………………………………………………………………… 137
　　　　例 4.3　个人档案文件管理 ex_43 ………………………………… 137
　　二、实验题 ……………………………………………………………… 147

　　　　实验 4.2　个人档案管理系统 …………………………… 147
第五章　高级汇编语言技术与连接技术 ………………………… 149
　5.1　高级汇编语言技术 …………………………………………… 149
　　一、示例 ………………………………………………………… 149
　　　　例 5.1　用宏和高级汇编技术实现 if 和 while 语句功能 ex-51 … 149
　　二、实验题 ……………………………………………………… 159
　　　　实验 5.1*　扩展 if 和 while 条件表达功能 ………………… 159
　5.2　连接技术 ……………………………………………………… 159
　　一、示例 ………………………………………………………… 159
　　　　例 5.2　可回卷的页显示 ex-52 ………………………………… 159
　　二、实验题 ……………………………………………………… 173
　　　　实验 5.2　菜单使用 ……………………………………………… 173
附录一　上机基本操作 ……………………………………………… 175
附录二　全屏幕编辑程序 WordStar ……………………………… 178
附录三　全屏幕编辑程序 pced …………………………………… 180
附录四　行编辑程序 EDLIN ……………………………………… 184
附录五　调试程序 DEBUG ………………………………………… 186
附录六　汇编程序出错信息 ………………………………………… 189
附录七　IBM-PC ASCII 码字符表 ……………………………… 194

第一章 实验的基本要求与方法

1.1 实验目的与要求

一、实验目的

学习程序设计的基本方法和技能,熟练掌握用汇编语言设计、编写、调试和运行程序的方法,为后续课程打下坚实的基础。

二、实验要求

1. 上机前要做好充分准备,包括程序框图、源程序清单、调试步骤、测试方法、对运行结果的分析等。

2. 上机时要遵守实验室的规章制度,爱护实验设备。要熟悉与实验有关的系统软件(如编辑程序、汇编程序、连接程序和调试程序等)的使用方法。在程序的调试过程中,有意识地学习及掌握 debug 程序的各种操作命令,以便掌握程序的调试方法及技巧。

为了更好地进行上机管理,要求用硬盘存储程序,并建立和使用子目录,以避免文件被别人删除。有关目录的操作命令见附录—4。此外,为便于统一管理硬盘中的文件,要求实验者按以下形式命名实验文件:

字母学号.扩充名

其中字母取 a~z 的 26 个英文字母,按实验顺序从 a 至 z 排列。如学号为 850431 学生的第二个实验程序所对应的文件名应为 b850431.asm。

3. 程序调试完后,须由实验辅导教师在机器上检查运行结果,经教师认可后的源程序可通过打印机输出,并请教师在程序清单上签字。每个实验完成后,应写出实验报告。实验报告的要求如下:

(1) 设计说明:用来说明程序的功能、结构。它包括:程序名、功能、原理及算法说明、程序及数据结构、主要符号名的说明等。

(2) 调试说明:便于学生总结经验提高编程及调试能力。它包括:调试情况,如上机时遇到的问题及解决办法,观察到的现象及其分析,对程序设计技巧的总结及分析等;程序的输出结果及对结果的分析;实验的心得体会,以及诸如调试日期、文件存放的软盘号等需要记录的信息。

(3) 使用说明:程序提供给用户使用时必须作出的说明。如:程序的使用方法,调用方式,操作步骤等;要求输入信息的类型及格式;出错信息的含义及程序的适用范围等。

(4) 程序框图。

(5) 经辅导教师签名后的程序清单。

1.2 实 验 方 法

有关汇编语言程序的上机过程请读者参阅清华大学出版社 1991 年出版的《IBM-PC 汇编语言程序设计》的 4.4 节。在这里,我们举例简要说明该过程以及程序的调试方法。

例 1.1   比较字符串 sample

试编写一程序:比较两个字符串 string1 和 string2 所含的字符是否相同。若相同则显示'Match',否则,显示'No match'。

我们可以用串比较指令来完成程序所要求的功能。上机过程如下:

1. 调用字处理程序 wordstar 建立 asm 文件

当然也可以用其他编辑程序如 pced 或行编辑程序 edlin 来建立源文件。

c＞ws

使用非文本文件方式(n 命令)建立以 sample.asm 为文件名的源文件如图 1.1 所示。然后用 CTRL K X 命令将文件存入磁盘,并使系统返回 DOS。

```
;PROGRAM TITLE GOES HERE－－Compare string
;*******************************************************
datarea segment              ;define data segment
  string1      db      'Move the cursor backward. '
  string2      db      'Move the cursor backward. '
;
  mess1        db      'Match. ',13,10,'$'
  mess2        db      'No match! ',13,10,'$'
datarea ends
;*******************************************************
prognam segment              ;define code segment
;------------------------------------------------------
main     proc    far
         assume cs:prognam,ds:datarea,es:datarea
start:                       ;starting execution address
;set up   stack for return
         push    ds          ;save old data segment
         sub     ax,ax       ;put zero in AX
         push    ax          ;save it on stack
;set DS register to current data segment
         mov     ax,datarea  ;datarea segment addr
         mov     ds,ax       ;  into DS register
         mov     es,ax       ;  into ES register
;MAIN PART OF PROGRAM GOES HERE
         lea     si,string1
         lea     di,string2
         cld
         mov     cx,25
         repz    cmpsb
```

```
            jz         match
            lea        dx,mess2
            jmp        short disp
match:
            lea        dx,mess1
disp:
            mov        ah,09
            int        21h
            ret                           ;return to DOS
main        endp                          ;end of main part of program
;——————————————————————————————
prognam    ends                           ;end of code segment
;**********************************************
            end        start              ;end assembly
```

图 1.1 例 1.1 的源文件 sample.asm

2. 用汇编程序 masm(或 asm)对源文件汇编产生目标文件 obj

C＞masm sample;
The IBM Personal Computer MACRO Assembler
Version 1.00 (C)Copyright IBM Corp 1981

Warning Severe
Errors Errors
  0   0

如汇编指示出错则需重新调用编辑程序修改错误,直至汇编通过为止。如调试时需要用 lst 文件,则应在汇编过程中建立该文件。

3. 用连接程序 link 产生执行文件 exe

C＞link sample;
IBM 5550 Multistation Linker 2.00
(C) Copyright IBM Corp. 1983

Warning：No STACK segment

There was 1 error detected.

4. 执行程序

可直接从 DOS 执行程序如下：
C＞sample
Match.

终端上已显示出程序的运行结果。为了调试程序的另一部分,可重新进编辑程序修改两个字符串的内容,使它们互不相同。如修改后的数据区为：

```
;**********************************************
datarea segment                ;define data segment
  string1         db       'Move the cursor backward.'
```

```
        string2       db          'Move the cursor forward.'
;
        mess1         db          'Match.',13,10,'$'
        mess2         db          'No match! ',13,10,'$'
datarea ends
;* * * * * * * * * * * * * * * * * * * * * * * * * * * * * * * * * *
```

然后,重新汇编、连接、执行,结果为:

C>sample
No match!

至此,程序已调试完毕,运行结果正确。

另一种调试程序的方法是使用 debug 程序。可调用如下:

C>debug sample.exe
—

此时,debug 已将执行程序装入内存,可直接用 g 命令运行程序。

—g
Match.

Program terminated normally

为调试程序的另一部分,可在 debug 中修改字符串内容。可先用 u 命令显示程序,以便了解指令地址。显示结果如图 1.2 所示。

```
—u
19F3:0000    1E              PUSH        DS
19F3:0001    2BC0            SUB         AX,AX
19F3:0003    50              PUSH        AX
19F3:0004    B8EE19          MOV         AX,19EE
19F3:0007    8ED8            MOV         DS,AX
19F3:0009    8EC0            MOV         ES,AX
19F3:000B    8D360000        LEA         SI,[0000]
19F3:000F    8D3E1900        LEA         DI,[0019]
19F3:0013    FC              CLD
19F3:0014    B91900          MOV         CX,0019
19F3:0017    F3              REPZ
19F3:0018    A6              CMPSB
19F3:0019    7406            JZ          0021
19F3:001B    8D163B00        LEA         DX,[003B]
19F3:001F    EB04            JMP         0025
—u
19F3:0021    8D163200        LEA         DX,[0032]
19F3:0025    B409            MOV         AH,09
19F3:0027    CD21            INT         21
19F3:0029    CB              RETF
19F3:002A    FF7501          PUSH        [DI+01]
19F3:002D    40              INC         AX
19F3:002E    5A              POP         DX
19F3:002F    22C2            AND         AL,DL
19F3:0031    50              PUSH        AX
19F3:0032    807EDC20        CMP         BYTE PTR [BP-24],20
19F3:0036    B0FF            MOV         AL,FF
19F3:0038    7201            JB          003B
```

```
19F3:003A  40              INC    AX
19F3:003B  5A              POP    DX
19F3:003C  22C2            AND    AL,DL
19F3:003E  50              PUSH   AX
19F3:003F  80F920          CMP    CL,20
```

图 1.2　例 1.1 用 debug 调试时，u 命令的显示情况

将断点设置在程序的主要部分运行以前。

—g0b

```
AX=19EE  BX=0000  CX=007A  DX=0000  SP=FFFC  BP=0000  SI=0000  DI=0000
DS=19EE  ES=19EE  SS=19EE  CS=19F3  IP=000B   NV UP DI PL ZR NA PE NC
19F3:000B  8D360000        LEA    SI,[0000]                     DS:0000=6F4D
```

根据其中指示的 ds 寄存器内容查看数据段的情况如下：

—d0

```
19EE:0000  4D 6F 76 65 20 74 68 65-20 63 75 72 73 6F 72 20   Move the cursor
19EE:0010  62 61 63 6B 77 61 72 64-2E 4D 6F 76 65 20 74 68   backward.Move th
19EE:0020  65 20 63 75 72 73 6F 72-20 62 61 63 6B 77 61 72   e cursor backwar
19EE:0030  64 2E 4D 61 74 63 68 2E-0D 0A 24 4E 6F 20 6D 61   d.Match...$No ma
19EE:0040  74 63 68 21 0D 0A 24 00-00 00 00 00 00 00 00 00   tch!..$.........
19EE:0050  1E 2B C0 50 B8 EE 19 8E-D8 8E C0 8D 36 00 00 8D   .+@P8n..X.@.6...
19EE:0060  3E 19 00 FC B9 19 00 F3-A6 74 06 8D 16 3B 00 EB   >..|9..s&t...;.k
19EE:0070  04 8D 16 32 00 B4 09 CD-21 CB FF 75 01 40 5A 22   ...2.4.M!K.u.@Z"
```

可用 e 命令修改数据区的字符串，操作如下：

—e29

```
19EE:0029  62.66    61.6f    63.72    6B.77    77.61    61.72    72.64
19EE:0030  64.2e    2E.20
```

再次用 d 命令查看修改结果。

—d0

```
19EE:0000  4D 6F 76 65 20 74 68 65-20 63 75 72 73 6F 72 20   Move the cursor
19EE:0010  62 61 63 6B 77 61 72 64-2E 4D 6F 76 65 20 74 68   backward.Move th
19EE:0020  65 20 63 75 72 73 6F 72-20 66 6F 72 77 61 72 64   e cursor forward
19EE:0030  2E 20 4D 61 74 63 68 2E-0D 0A 24 4E 6F 20 6D 61   . Match...$No ma
19EE:0040  74 63 68 21 0D 0A 24 00-00 00 00 00 00 00 00 00   tch!..$.........
19EE:0050  1E 2B C0 50 B8 EE 19 8E-D8 8E C0 8D 36 00 00 8D   .+@P8n..X.@.6...
19EE:0060  3E 19 00 FC B9 19 00 F3-A6 74 06 8D 16 3B 00 EB   >..|9..s&t...;.k
19EE:0070  04 8D 16 32 00 B4 09 CD-21 CB FF 75 01 40 5A 22   ...2.4.M!K.u.@Z"
```

用 g 命令运行程序，结果为：

—g
No match!

Program terminated normally

用 q 命令退出 debug。

—q

至此，程序已调试完毕。为了进一步说明 debug 命令的使用方法，我们再次重复上述程序的调试过程，只是使用 e、a 和 f 命令来修改数据区的内容而已。必须注意，由于在用 debug 调试程序时，只能修改当时有关的内存单元内容，因此重新用 debug 装入执行程序

时,仍是原来在磁盘中的文件内容。操作如下：
C＞debug sample.exe
-g0b

```
AX=19EE    BX=0000    CX=007A    DX=0000    SP=FFFC    BP=0000    SI=0000    DI=0000
DS=19EE    ES=19EE    SS=19EE    CS=19F3    IP=000B    NV UP DI PL ZR NA PE NC
19F3:000B  8D360000       LEA       SI,[0000]                    DS:0000=6F4D
```

-d0
```
19EE:0000   4D 6F 76 65 20 74 68  65-20  63 75 72 73 6F 72 20    Move the cursor
19EE:0010   62 61 63 6B 77 61 72  64-2E  4D 6F 76 65 20 74 68    backward.Move th
19EE:0020   65 20 63 75 72 73 6F  72-20  62 61 63 6B 77 61 72    e cursor backwar
19EE:0030   64 2E 4D 61 74 63 68  21-0D  0A 24 4E 6F 20 6D 61    d.Match...$No ma
19EE:0040   74 63 68 21 0D 0A 24  00-00  00 00 00 00 00 00 00    tch!..$.........
19EE:0050   1E 2B C0 50 B8 EE 19  8E-D8  8E C0 8D 36 00 00 8D    .+@P8n..X.@.6...
19EE:0060   3E 19 00 FC B9 19 00  F3-A6  74 06 8D 16 3B 00 EB    >..|9..s&t..;.k
19EE:0070   04 8D 16 32 00 B4 09  CD-21  CB FF 75 01 40 5A 22    ...2.4.M!K.u.@Z"
```

-e29   'forward.' 20

-d0
```
19EE:0000   4D 6F 76 65 20 74 68  65-20  63 75 72 73 6F 72 20    Move the cursor
19EE:0010   62 61 63 6B 77 61 72  64-2E  4D 6F 76 65 20 74 68    backward.Move th
19EE:0020   65 20 63 75 72 73 6F  72-20  66 6F 72 77 61 72 64    e cursor forward
19EE:0030   2E 20 4D 61 74 63 68  21-0D  0A 24 4E 6F 20 6D 61    . Match...$No ma
19EE:0040   74 63 68 21 0D 0A 24  00-00  00 00 00 00 00 00 00    tch!..$.........
19EE:0050   1E 2B C0 50 B8 EE 19  8E-D8  8E C0 8D 36 00 00 8D    .+@P8n..X.@.6...
19EE:0060   3E 19 00 FC B9 19 00  F3-A6  74 06 8D 16 3B 00 EB    >..|9..s&t..;.k
19EE:0070   04 8D 16 32 00 B4 09  CD-21  CB CC CC CC CC CC CC    ...2.4.M!KLLLLLL
```

-g
No match!

Program terminated normally

可见这种 e 命令的方式避免使用 ASCII 码进入,对用户是比较方便的。其中最后一个 20 是空格键的 ASCII 码,以补足原来的字节数。

也可使用 a 命令把数据区的内容恢复原状,具体如下：

-a19ee:29
19EE:0029   db   'backward.'
19EE:0032

-d0
```
19EE:0000   4D 6F 76 65 20 74 68  65-20  63 75 72 73 6F 72 20    Move the cursor
19EE:0010   62 61 63 6B 77 61 72  64-2E  4D 6F 76 65 20 74 68    backward.Move th
19EE:0020   65 20 63 75 72 73 6F  72-20  62 61 63 6B 77 61 72    e cursor backwar
19EE:0030   64 2E 4D 61 74 63 68  21-0D  0A 24 4E 6F 20 6D 61    d.Match...$No ma
19EE:0040   74 63 68 21 0D 0A 24  00-00  00 00 00 00 00 00 00    tch!..$.........
19EE:0050   1E 2B C0 50 B8 EE 19  8E-D8  8E C0 8D 36 00 00 8D    .+@P8n..X.@.6...
19EE:0060   3E 19 00 FC B9 19 00  F3-A6  74 06 8D 16 3B 00 EB    >..|9..s&t..;.k
19EE:0070   04 8D 16 32 00 B4 09  CD-21  CB 6F 72 77 61 72 64    ...2.4.M!Korward
```

-g
Match.
AX=0924   BX=0000   CX=0000   DX=0032   SP=FFFC   BP=FFFF   SI=0019   DI=0032

```
DS=19EE    ES=19EE    SS=19EE    CS=19EE    IP=0096    NV UP DI NG NZ AC PE NC
19EE:0096 CC                         INT        3
-q
```

由于 a 命令是汇编命令,因此信息是用汇编格式进入的。如果修改的是程序中的语句,方法也是相同的,下面我们还会看到这类的操作。现在再看一下用 f 命令修改数据区的方法。

```
C>debug sample.exe
-g0b

AX=19EE    BX=0000    CX=007A    DX=0000    SP=FFFC    BP=0000    SI=0000    DI=0000
DS=19EE    ES=19EE    SS=19EE    CS=19F3    IP=000B    NV UP DI PL ZR NA PE NC
19F3:000B 8D360000         LEA       SI,[0000]                  DS:0000=6F4D
-d0
19EE:0000   4D 6F 76 65 20 74 68 65-20 63 75 72 73 6F 72 20   Move the cursor
19EE:0010   62 61 63 6B 77 61 72 64-2E 4D 6F 76 65 20 74 68   backward.Move th
19EE:0020   65 20 63 75 72 73 6F 72-20 62 61 63 6B 77 61 72   e cursor backwar
19EE:0030   64 2E 4D 61 74 63 68 2E-0D 0A 24 4E 6F 20 6D 61   d.Match...$No ma
19EE:0040   74 63 68 21 0D 0A 24 00-00 00 00 00 00 00 00 00   tch!..$.........
19EE:0050   1E 2B C0 50 B8 EE 19 8E-D8 8E C0 8D 36 00 00 8D   .+@P8n..X.@.6...
19EE:0060   3E 19 00 FC B9 19 00 F3-A6 74 06 8D 16 3B 00 EB   >..│9..s&t...;.k
19EE:0070   04 8D 16 32 00 B4 09 CD-21 CB FF 75 01 40 5A 22   ...2.4.M!K.u.@Z"

-f29 1 9 'forward.'20
-d0
19EE:0000   4D 6F 76 65 20 74 68 65-20 63 75 72 73 6F 72 20   Move the cursor
19EE:0010   62 61 63 6B 77 61 72 64-2E 4D 6F 76 65 20 74 68   backward.Move th
19EE:0020   65 20 63 75 72 73 6F 72-20 66 6F 72 77 61 72 64   e cursor forward
19EE:0030   2E 20 4D 61 74 63 68 2E-0D 0A 24 4E 6F 20 6D 61   . Match...$No ma
19EE:0040   74 63 68 21 0D 0A 24 00-00 00 00 00 00 00 00 00   tch!..$.........
19EE:0050   1E 2B C0 50 B8 EE 19 8E-D8 8E C0 8D 36 00 00 8D   .+@P8n..X.@.6...
19EE:0060   3E 19 00 FC B9 19 00 F3-A6 74 06 8D 16 3B 00 EB   >..│9..s&t...;.k
19EE:0070   04 8D 16 32 00 B4 09 CD-21 CB FF 75 01 40 5A 22   ...2.4.M!K.u.@Z"

-g
No match!

Program terminated normally
-q
```

f 命令中的 29 为所修改区的首地址,19 表示需要修改的长度为 9 个字节。

为进一步说明程序的调试过程,现假设程序编制错误:在源文件中把 jz match 改为 jnz match。该程序经汇编、连接后,进入 debug 调试如下:

```
C>debug sample.exe
-g
No match!

Program terminated normally
```

结果是错误的(因源文件中两个字符串是相同的)。为检查程序的错误,将断点设在比较串之后。

```
-g19
```

```
AX=19EE  BX=0000  CX=0000  DX=0000  SP=FFFC  BP=0000  SI=0019  DI=0032
DS=19EE  ES=19EE  SS=19EE  CS=19F3  IP=0019           NV UP DI PL ZR NA PE NC
19F3:0019 7506              JNZ        0021
```

此时零标志为 ZR，即 ZF=1，即表示比较结果相等，说明比较结果是正确的。现在可用 p 命令再执行一条指令以观察指令的转向。

　　—p

```
AX=19EE  BX=0000  CX=0000  DX=0000  SP=FFFC  BP=0000  SI=0019  DI=0032
DS=19EE  ES=19EE  SS=19EE  CS=19F3  IP=001B           NV UP DI PL ZR NA PE NC
19F3:001B 8D163B00          LEA        DX,[003B]                DS:003B=6F4E
```

为查到 003B 单元的内容，可查数据区如下：

　　—d0

```
19EE:0000  4D 6F 76 65 20 74 68 65-20 63 75 72 73 6F 72 20   Move the cursor 
19EE:0010  62 61 63 6B 77 61 72 64-2E 4D 6F 76 65 20 74 68   backward.Move th
19EE:0020  65 20 63 75 72 73 6F 72-20 62 61 63 6B 77 61 72   e cursor backwar
19EE:0030  64 2E 4D 61 74 63 68 68-2E 0D 0A 24 4E 6F 20 6D   d.Match...$No m
19EE:0040  61 74 63 68 21 0D 0A 24-00 00 00 00 00 00 00 00   atch!..$........
19EE:0050  1E 2B C0 50 B8 EE 19 8E-D8 8E C0 8D 36 00 00 8D   .+@P8n..X.@.6...
19EE:0060  3E 19 00 FC B9 19 00 F3-A6 75 06 8D 16 3B 00 EB   >..|9.s&u...;.k
19EE:0070  04 8D 16 32 00 B4 09 CD-21 CB FF 75 01 40 5A 22   ...2.4.M!K.u.@Z"
```

可见 003B 单元的内容为 4E，即 N 的 ASCII 码，后面跟的是 No match!，这说明 jnz 指令使用错误，应该改为 jz match。可用 a 命令修改，并用 u 命令检查修改结果。运行结果说明程序修改正确。

```
—a19
19F3:0019           jz         0021
19F3:001B
—u0
19F3:0000  1E              PUSH       DS
19F3:0001  2BC0            SUB        AX,AX
19F3:0003  50              PUSH       AX
19F3:0004  B8EE19          MOV        AX,19EE
19F3:0007  8ED8            MOV        DS,AX
19F3:0009  8EC0            MOV        ES,AX
19F3:000B  8D360000        LEA        SI,[0000]
19F3:000F  8D3E1900        LEA        DI,[0019]
19F3:0013  FC              CLD
19F3:0014  B91900          MOV        CX,0019
19F3:0017  F3              REPZ
19F3:0018  A6              CMPSB
19F3:0019  7406            JZ         0021
19F3:001B  8D163B00        LEA        DX,[003B]
19F3:001F  EB04            JMP        0025
—rip
IP 001B
:0
—g
Match.
```

```
AX=0924  BX=0000  CX=0000  DX=0032  SP=FFFC  BP=FFFF  SI=0019  DI=0032
DS=19EE  ES=19EE  SS=19EE  CS=19EE  IP=0096  NV UP DI NG NZ AC PE NC
19EE:0096 CC              INT      3
-q
```

在这里应该注意,在使用 a 命令修改数据区时,必须给出数据段的地址,而在修改程序区时,由于 a 命令的缺省段为代码段,所以直接给出偏移地址就可以了。

在调试过程中,也可以用 t 命令逐条跟踪程序的执行。下面列出断点停在 0b 后,用 f 命令修改数据区中字符串的内容,然后用 t 命令逐条执行指令的情况。

```
-f29 1 9 'forward.' 20
-d0
19EE:0000  4D 6F 76 65 20 74 68 65-20 63 75 72 73 6F 72 20   Move the cursor
19EE:0010  62 61 63 6B 77 61 72 64-2E 4D 6F 76 65 20 74 68   backward.Move th
19EE:0020  65 20 63 75 72 73 6F 72-20 66 6F 72 77 61 72 64   e cursor forward
19EE:0030  2E 20 4D 61 74 63 68 2E-0D 0A 24 4E 6F 20 6D 61   . Match...$No ma
19EE:0040  74 63 68 21 0D 0A 24 00-00 00 00 00 00 00 00 00   tch!..$.........
19EE:0050  1E 2B C0 50 B8 EE 19 8E-D8 8E C0 8D 36 00 00 8D   .+@P8n..X.@.6...
19EE:0060  3E 19 00 FC B9 19 00 F3-A6 74 06 8D 16 3B 00 EB   >..|9..s&t...;.k
19EE:0070  04 8D 16 32 00 B4 09 CD-21 CB FF 75 01 40 5A 22   ...2.4.M!K.u.@Z"
-t

AX=19EE  BX=0000  CX=007A  DX=0000  SP=FFFC  BP=0000  SI=0000  DI=0000
DS=19EE  ES=19EE  SS=19EE  CS=19F3  IP=000F  NV UP DI PL ZR NA PE NC
19F3:000F 8D3E1900        LEA      DI,[0019]                  DS:0019=6F4D
-t

AX=19EE  BX=0000  CX=007A  DX=0000  SP=FFFC  BP=0000  SI=0000  DI=0019
DS=19EE  ES=19EE  SS=19EE  CS=19F3  IP=0013  NV UP DI PL ZR NA PE NC
19F3:0013 FC              CLD
-t

AX=19EE  BX=0000  CX=007A  DX=0000  SP=FFFC  BP=0000  SI=0000  DI=0019
DS=19EE  ES=19EE  SS=19EE  CS=19F3  IP=0014  NV UP DI PL ZR NA PE NC
19F3:0014 B91900          MOV      CX,0019
-t

AX=19EE  BX=0000  CX=0019  DX=0000  SP=FFFC  BP=0000  SI=0000  DI=0019
DS=19EE  ES=19EE  SS=19EE  CS=19F3  IP=0017  NV UP DI PL ZR NA PE NC
19F3:0017 F3              REPZ
19F3:0018 A6              CMPSB
-t

AX=19EE  BX=0000  CX=0018  DX=0000  SP=FFFC  BP=0000  SI=0001  DI=001A
DS=19EE  ES=19EE  SS=19EE  CS=19F3  IP=0017  NV UP DI PL ZR NA PE NC
19F3:0017 F3              REPZ
19F3:0018 A6              CMPSB
-t

AX=19EE  BX=0000  CX=0017  DX=0000  SP=FFFC  BP=0000  SI=0002  DI=001B
DS=19EE  ES=19EE  SS=19EE  CS=19F3  IP=0017  NV UP DI PL ZR NA PE NC
19F3:0017 F3              REPZ
19F3:0018 A6              CMPSB
-t

AX=19EE  BX=0000  CX=0016  DX=0000  SP=FFFC  BP=0000  SI=0003  DI=001C
```

```
DS=19EE    ES=19EE    SS=19EE    CS=19F3    IP=0017    NV UP DI PL ZR NA PE NC
19F3:0017 F3                     REPZ
19F3:0018 A6                     CMPSB
-t

AX=19EE    BX=0000    CX=0015    DX=0000    SP=FFFC    BP=0000    SI=0004    DI=001D
DS=19EE    ES=19EE    SS=19EE    CS=19F3    IP=0017    NV UP DI PL ZR NA PE NC
19F3:0017 F3                     REPZ
19F3:0018 A6                     CMPSB
-t

AX=19EE    BX=0000    CX=0014    DX=0000    SP=FFFC    BP=0000    SI=0005    DI=001E
DS=19EE    ES=19EE    SS=19EE    CS=19F3    IP=0017    NV UP DI PL ZR NA PE NC
19F3:0017 F3                     REPZ
19F3:0018 A6                     CMPSB
-t

AX=19EE    BX=0000    CX=0013    DX=0000    SP=FFFC    BP=0000    SI=0006    DI=001F
DS=19EE    ES=19EE    SS=19EE    CS=19F3    IP=0017    NV UP DI PL ZR NA PE NC
19F3:0017 F3                     REPZ
19F3:0018 A6                     CMPSB
-t

AX=19EE    BX=0000    CX=0012    DX=0000    SP=FFFC    BP=0000    SI=0007    DI=0020
DS=19EE    ES=19EE    SS=19EE    CS=19F3    IP=0017    NV UP DI PL ZR NA PE NC
19F3:0017 F3                     REPZ
19F3:0018 A6                     CMPSB
-t

AX=19EE    BX=0000    CX=0011    DX=0000    SP=FFFC    BP=0000    SI=0008    DI=0021
DS=19EE    ES=19EE    SS=19EE    CS=19F3    IP=0017    NV UP DI PL ZR NA PE NC
19F3:0017 F3                     REPZ
19F3:0018 A6                     CMPSB
-t

AX=19EE    BX=0000    CX=0010    DX=0000    SP=FFFC    BP=0000    SI=0009    DI=0022
DS=19EE    ES=19EE    SS=19EE    CS=19F3    IP=0017    NV UP DI PL ZR NA PE NC
19F3:0017 F3                     REPZ
19F3:0018 A6                     CMPSB
-t

AX=19EE    BX=0000    CX=000F    DX=0000    SP=FFFC    BP=0000    SI=000A    DI=0023
DS=19EE    ES=19EE    SS=19EE    CS=19F3    IP=0017    NV UP DI PL ZR NA PE NC
19F3:0017 F3                     REPZ
19F3:0018 A6                     CMPSB
-t

AX=19EE    BX=0000    CX=000E    DX=0000    SP=FFFC    BP=0000    SI=000B    DI=0024
DS=19EE    ES=19EE    SS=19EE    CS=19F3    IP=0017    NV UP DI PL ZR NA PE NC
19F3:0017 F3                     REPZ
19F3:0018 A6                     CMPSB
-t

AX=19EE    BX=0000    CX=000D    DX=0000    SP=FFFC    BP=0000    SI=000C    DI=0025
DS=19EE    ES=19EE    SS=19EE    CS=19F3    IP=0017    NV UP DI PL ZR NA PE NC
19F3:0017 F3                     REPZ
19F3:0018 A6                     CMPSB
-t
```

```
AX=19EE    BX=0000    CX=000C    DX=0000    SP=FFFC    BP=0000    SI=000D    DI=0026
DS=19EE    ES=19EE    SS=19EE    CS=19F3    IP=0017    NV UP DI PL ZR NA PE NC
19F3:0017 F3                     REPZ
19F3:0018 A6                     CMPSB
-t

AX=19EE    BX=0000    CX=000B    DX=0000    SP=FFFC    BP=0000    SI=000E    DI=0027
DS=19EE    ES=19EE    SS=19EE    CS=19F3    IP=0017    NV UP DI PL ZR NA PE NC
19F3:0017 F3                     REPZ
19F3:0018 A6                     CMPSB
-t

AX=19EE    BX=0000    CX=000A    DX=0000    SP=FFFC    BP=0000    SI=000F    DI=0028
DS=19EE    ES=19EE    SS=19EE    CS=19F3    IP=0017    NV UP DI PL ZR NA PE NC
19F3:0017 F3                     REPZ
19F3:0018 A6                     CMPSB
-t

AX=19EE    BX=0000    CX=0009    DX=0000    SP=FFFC    BP=0000    SI=0010    DI=0029
DS=19EE    ES=19EE    SS=19EE    CS=19F3    IP=0017    NV UP DI PL ZR NA PE NC
19F3:0017 F3                     REPZ
19F3:0018 A6                     CMPSB
-t

AX=19EE    BX=0000    CX=0008    DX=0000    SP=FFFC    BP=0000    SI=0011    DI=002A
DS=19EE    ES=19EE    SS=19EE    CS=19F3    IP=0019    NV UP DI NG NZ AC PE CY
19F3:0019 7406                   JZ         0021
-t

AX=19EE    BX=0000    CX=0008    DX=0000    SP=FFFC    BP=0000    SI=0011    DI=002A
DS=19EE    ES=19EE    SS=19EE    CS=19F3    IP=001B    NV UP DI NG NZ AC PE CY
19F3:001B 8D163B00               LEA        DX,[003B]                        DS:003B=6F4E
-t

AX=19EE    BX=0000    CX=0008    DX=003B    SP=FFFC    BP=0000    SI=0011    DI=002A
DS=19EE    ES=19EE    SS=19EE    CS=19F3    IP=001F    NV UP DI NG NZ AC PE CY
19F3:001F EB04                   JMP        0025
-t

AX=19EE    BX=0000    CX=0008    DX=003B    SP=FFFC    BP=0000    SI=0011    DI=002A
DS=19EE    ES=19EE    SS=19EE    CS=19F3    IP=0025    NV UP DI NG NZ AC PE CY
19F3:0025 B409                   MOV        AH,09
-t

AX=09EE    BX=0000    CX=0008    DX=003B    SP=FFFC    BP=0000    SI=0011    DI=002A
DS=19EE    ES=19EE    SS=19EE    CS=19F3    IP=0027    NV UP DI NG NZ AC PE CY
19F3:0027 CD21                   INT        21
-g
No match!

Program terminated normally
-q
```

从这一过程可清楚地看出每次比较的结果,一旦比较不等,则立即从串指令退出,执行下面的指令。应该注意,如果遇到系统功能调用,则不能再使用 t 或 p 命令跟踪,而应该用断点停在功能调用完成之后,然后再接着跟踪。本例中,由于不需要再跟踪,所以直接用 g 命令运行到程序结束。

前面已经提到，debug 调试期间所修改的数据段或代码段的内容只是修改了内存中的内容，而磁盘文件中的内容并未修改。如果你的执行文件是.com，则可在 debug 中用 n、w 命令直接把经修改后的内存单元中的内容存入磁盘，但是.exe 文件则不允许这样做，因此，应该重新进入编辑程序，根据调试结果把源文件修改正确，经汇编、连接、执行检查正确后再保存下来。

第二章 程序的基本结构练习

2.1 循环程序设计

一、示例

例 2.1 表格查找 tabsrch

仓库管理中,总共存有有关库存品的编号、名称、数量、价格等情况的表格,根据用户提供的编号可以找到有关材料。假设表格中共有 6 种库存品,表格的格式为:

```
STOKTAB    DB    '05', 'Excavators'
                 '08', 'lifters    '
                  .
                  .
                  .
```

试编写一程序,根据用户提供的编号在终端上显示其名称。

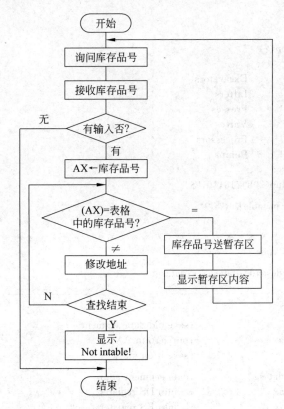

图 2.1 例 2.1 的程序框图

根据题目的要求,程序可由三部分组成:
(1) 输入:接收用户键入的库存品号;
(2) 查表:根据给定编号在表格中查找所要求的库存品名称;
(3) 输出:显示找到的库存品名称。

其中,第二部分是程序的主体,显然它可以使用循环结构。由于在给定的表格中,编号及库存品名所占的字节数都是相等的,因此在查找时,只要根据编号的地址就可逐项查找。循环的结束条件应该是在找到用户所指定的项时就可退出循环,但还必须考虑在表格中未查到所需编号的特殊情况。根据以上考虑可画出程序框图如图 2.1 所示。根据框图编制出如图 2.2 所示的程序。图 2.3 给出了程序的运行情况。可以看出,如用户给出的编号可在表格中查到,则显示出相应的库存品名。如用户不再需要查找,只需按一下回车键就可结束程序。如表格中并无用户给出的编号,则显示 Not in table! 后结束程序。

```
;PROGRAM TITLE GOES HERE－－tabsrch
;Table search
;＊＊＊＊＊＊＊＊＊＊＊＊＊＊＊＊＊＊＊＊＊＊＊＊＊＊＊＊＊＊＊＊＊＊＊＊
datasg    segment   para      'data'
mess1     db        'stock nember? ',13,10,'$'
;
stoknin   label     byte
  max     db        3
  act     db        ?
  stokn   db        3 dup(?)
;
stoktab   db        '05','    Excavators      '
          db        '08','    Lifters         '
          db        '09','    Presses         '
          db        '12','    Valves          '
          db        '23','    Processors      '
          db        '27','    Pumps           '
;
descrn    db        14 dup(20h),13,10,'$'
mess      db        'Not in table! ','$'
datasg    ends
;＊＊＊＊＊＊＊＊＊＊＊＊＊＊＊＊＊＊＊＊＊＊＊＊＊＊＊＊＊＊＊＊＊＊＊＊
codesg    segment   para 'code'
          assume    cs:codesg,ds:datasg,es:datasg
;－－－－－－－－－－－－－－－－－－－－－－－－－－－－－－－－－－
main      proc      far
          push      ds                ;save old data segment
          sub       ax,ax             ;put zero in AX
          push      ax                ;save it on stack
          mov       ax,datasg         ;data segment addr
          mov       ds,ax             ;   into DS register
          mov       es,ax             ;   into ES register
;MAIN PART OF PROGRAM GOES HERE
start:
          lea       dx,mess1          ;Prompt for stock number
```

```
            mov     ah,09
            int     21h
            lea     dx,stoknin
            mov     ah,0ah
            int     21h
            cmp     act,0
            je      exit
            mov     al,stokn            ;Get stock#
            mov     ah,stokn+1
            mov     cx,06               ;No. of entries
            lea     si,stoktab          ;Init'ze table address
a20:
            cmp     ax,word ptr[si]     ;Stock# : table
            je      a30                 ;Equal — exit
            add     si,14               ;Not equal — increment
            loop    a20
            lea     dx,mess             ;Not in table
            mov     ah,09
            int     21h
            jmp     exit
a30:
            mov     cx,07               ;Length of descr'n
            lea     di,descrn           ;Addr of descr'n
            rep     movsw
;
            lea     dx,descrn
            mov     ah,09
            int     21h
            jmp     start
exit:
            ret                         ;return to DOS
main        endp
;————————————————————————————————————————
codesg      ends                        ;end of code segment
;************************************************
            end     main                ;end assembly
```

图 2.2 例 2.1 的程序清单

C>tsrch
stock nember?
23   Processors
stock nember?
27   Pumps
stock nember?
05   Excavators
stock nember?

C>tsrch
stock nember?
09   Presses
stock nember?

```
08  Lifters
stock nember?
12  Valves
stock nember?
Not in table!
```

图 2.3 例 2.1 的运行情况

例 2.2 建立学生名次表 rank

以 grade 为首地址的 10 个字的数组中保存了学生的成绩，其中 grade+i 保存学号为

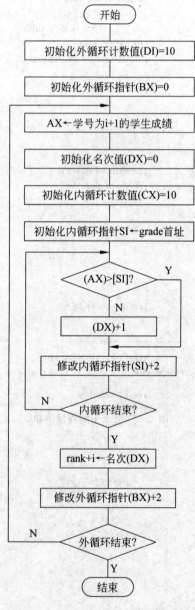

图 2.4 例 2.2 的程序框图

i+1 的学生的成绩。要求建立一个 10 个字的 rank 数组,并根据 grade 中的学生成绩将学生名次填入 rank 数组中,其中 rank+i 的内容是学号为 i+1 学生的名次(提示:一个学生的名次等于成绩高于该学生的人数加 1)。

本题可以用两重循环结构来实现。内层对应于每个学生的名次计算,外层则解决所有学生的名次计算。由于数组的长度是确定的,因此内、外层均可用计数值来控制循环的结束。在设计多重循环结构时,必须注意内层循环结束后的重新初始化问题。由于本例所用寄存器较多,现将寄存器的分配情况说明如下:

  AX 存放当前被测学生的成绩;
  BX 存放当前被测学生的相对地址指针;
  CX 内循环计数值;
  DX 当前被测学生的名次计数值;
  SI 内循环测试时的地址指针;
  DI 外循环计数值。

根据以上考虑,画出程序框图如图 2.4 所示。程序清单如图 2.5 所示。程序运行情况如图 2.6 所示。这里必须使用 debug 来查看程序的运行结果,可以看出当程序结束时,rank 数组已存放了学生的名次。

```
;PROGRAM TITLE GOES HERE——Rank
;*********************************************
datarea    segment                  ;define data segment
  grade          dw        88,75,95,63,98,78,87,73,90,60
  rank           dw        10 dup(?)
datarea ends
;*********************************************
prognam segment                     ;define code segment
;————————————————————————————————————
main       proc       far           ;main part of program
           assume cs:prognam,ds:datarea
start:                               ;starting execution address
;set up stack for return
           push       ds            ;save old data segment
           sub        ax,ax         ;put zero in AX
           push       ax            ;save it on stack
;set DS register to current data segment
           mov        ax,datarea    ;datarea segment addr
           mov        ds,ax         ;  into DS register
;MAIN PART OF PROGRAM GOES HERE
           mov        di,10
           mov        bx,0
loop:
           mov        ax,grade[bx]
           mov        dx,0
           mov        cx,10
           lea        si,grade
next:
           cmp        ax,[si]
           jg         no_count
```

```
                inc     dx
no-count:
                add     si,2
                loop    next
                mov     rank[bx],dx
                add     bx,2
                dec     di
                jne     loop
                ret                             ;return to DOS
main    endp                                    ;end of main part of program
;------------------------------------------------------------------
prognam ends                                    ;end of code segment
;* * * * * * * * * * * * * * * * * * * * * * * * * * * * * * * * *
                end     start                   ;end assembly
```

图 2.5 例 2.2 的程序清单

```
C:\ >debug rank.exe
-u
19F3:0000   1E              PUSH    DS
19F3:0001   2BC0            SUB     AX,AX
19F3:0003   50              PUSH    AX
19F3:0004   B8F019          MOV     AX,19F0
19F3:0007   8ED8            MOV     DS,AX
19F3:0009   BF0A00          MOV     DI,000A
19F3:000C   BB0000          MOV     BX,0000
19F3:000F   8B870000        MOV     AX,[BX+0000]
19F3:0013   BA0000          MOV     DX,0000
19F3:0016   B90A00          MOV     CX,000A
19F3:0019   8D360000        LEA     SI,[0000]
19F3:001D   3B04            CMP     AX,[SI]
19F3:001F   7F01            JG      0022
-u
19F3:0021   42              INC     DX
19F3:0022   83C602          ADD     SI,+02
19F3:0025   E2F6            LOOP    001D
19F3:0027   89971400        MOV     [BX+0014],DX
19F3:002B   83C302          ADD     BX,+02
19F3:002E   4F              DEC     DI
19F3:002F   75DE            JNZ     000F
19F3:0031   CB              RETF
19F3:0032   5A              POP     DX
19F3:0033   22C2            AND     AL,DL
19F3:0035   50              PUSH    AX
19F3:0036   80F93B          CMP     CL,3B
19F3:0039   B0FF            MOV     AL,FF
19F3:003B   7501            JNZ     003E
19F3:003D   40              INC     AX
19F3:003E   5A              POP     DX
19F3:003F   22C2            AND     AL,DL
-g09
AX=19F0   BX=0000   CX=0062   DX=0000   SP=FFFC   BP=0000   SI=0000   DI=0000
DS=19F0   ES=19E0   SS=19F0   CS=19F3   IP=0009   NV UP DI PL ZR NA PE NC
19F3:0009 BF0A00            MOV     DI,000A
```

```
—d0
19F0:0000    58 00 4B 00 5F 00 3F 00-62 00 4E 00 57 00 49 00    X.K._.?.b.N.W.I.
19F0:0010    5A 00 3C 00 00 00 00 00-00 00 00 00 00 00 00 00    Z.<.............
19F0:0020    00 00 00 00 00 00 00 00-00 00 00 00 00 00 00 00    ................
19F0:0030    1E 2B C0 50 B8 F0 19 8E-D8 BF 0A 00 BB 00 00 8B    .+@P8p..X?..;...
19F0:0040    87 00 00 BA 00 00 B9 0A-00 8D 36 00 00 3B 04 7F    ..........9..6.;.
19F0:0050    01 42 83 C6 02 E2 F6 89-97 14 00 83 C3 02 4F 75    .B.F.bv.....C.Ou
19F0:0060    DE CB 5A 22 C2 50 80 F9-3B B0 FF 75 01 40 5A 22    ^KZ"BP.y;0.u.@Z"
19F0:0070    C2 50 80 7E DC 20 B0 FF-72 01 40 5A 22 C2 50 80    BP.~\ 0.r.@Z"BP.
—g31
AX=003C   BX=0014   CX=0000   DX=000A   SP=FFFC   BP=0000   SI=0014   DI=0000
DS=19F0   ES=19E0   SS=19F0   CS=19F3   IP=0031   NV UP DI PL ZR NA PE NC
19F3:0031 CB         RETF
—d0
19F0:0000    58 00 4B 00 5F 00 3F 00-62 00 4E 00 57 00 49 00    X.K._.?.b.N.W.I.
19F0:0010    5A 00 3C 00 00 00 07 00-02 00 09 00 01 00 06 00    Z.<.............
19F0:0020    05 00 08 00 03 00 0A 00-00 00 00 00 00 00 00 00    ................
19F0:0030    1E 2B C0 50 B8 F0 19 8E-D8 BF 0A 00 BB 00 00 8B    .+@P8p..X?..;...
19F0:0040    87 00 00 BA 00 00 B9 0A-00 8D 36 00 00 3B 04 7F    ..........9..6.;.
19F0:0050    01 42 83 C6 02 E2 F6 89-97 14 00 83 C3 02 4F 75    .B.F.bv.....C.Ou
19F0:0060    DE CB 5A 22 C2 50 80 F9-3B B0 FF 75 01 40 5A 22    ^KZ"BP.y;0.u.@Z"
19F0:0070    C2 50 80 7E DC 20 B0 FF-72 01 40 5A 22 C2 50 80    BP.~\ 0.r.@Z"BP.
—q
```

图 2.6 例 2.2 的运行情况

二、实验题

实验 2.1 用表格形式显示字符

1. 题目:用表格形式显示 ASCII 字符 SMASCII
2. 实验要求:

按 15 行×16 列的表格形式显示 ASCII 码为 10H～100H 的所有字符,即以行为主的顺序及 ASCII 码递增的次序依次显示对应的字符。每 16 个字符为一行,每行中的相邻两个字符之间用空白符(ASCII 为 0)隔开。

3. 提示:

(1) 显示每个字符可使用功能号为 02 的显示输出功能调用,使用方法如下:

```
    mov     ah,02h
    mov     dl,输出字符的ASCII码
    int     21h
```

本题中可把 dl 初始化为 10H,然后不断使其加 1(用 INC 指令)以取得下一个字符的 ASCII 码。

(2) 显示空白符时,用其 ASCII 码 0 置入 dl 寄存器。每行结束时,用显示回车(ASCII 为 0dh)和换行符(ASCII 为 0ah)来结束本行并开始下一行。

(3) 由于逐个显示相继的 ASCII 字符时,需要保存并不断修改 dl 寄存器的内容,而显示空白、回车、换行符时也需要使用 dl 寄存器,为此可使用堆栈来保存相继的 ASCII 字符。具体用法是:在显示空白或回车、换行符前用指令

$$\text{push} \quad \text{dx}$$

把 dl 的内容保存到堆栈中去。在显示空白或回车、换行符后用指令

$$\text{pop} \quad \text{dx}$$

恢复 dl 寄存器的原始内容。

实验 2.2　查找匹配字符串

1. 题目：查找匹配字符串 SEARCH
2. 实验要求：

程序接收用户键入的一个关键字以及一个句子。如果句子中不包含关键字则显示 'No match!'；如果句子中包含关键字则显示 'Match'，且把该字在句子中的位置用十六进制数显示出来。要求程序的执行过程如下：

Enter keyword:abc
Enter Sentence:We are studying abc.
Match at location:11 H of the sentence.
Enter Sentence:xyz,OK?
No match.
Enter Sentence:^C

3. 提示：

程序可由三部分组成：

(1) 输入关键字和一个句子，分别存入相应的缓冲区中。可用功能调用 0AH。

(2) 在句子中查找关键字。

① 关键字和句子中相应字段的比较可使用串比较指令。为此必须定义附加段，但附加段和数据段可定义为同一段，以便于串指令的使用。这样，相应的寄存器内容也有了确定的含义，即如下：

SI　　寄存器为关键字的指针；

DI　　寄存器为句子中正相比较的字段的指针；

CX　　寄存器存放关键字的字母个数（长度）。

② 整个句子和关键字的比较过程可以用一个循环结构来完成。循环次数为：

（句子长度－关键字长度）＋1

在计算循环次数时，如遇到句子长度小于关键字长度的情况则应转向显示 'No match!'。循环中还需要用到 BX 寄存器，它用来保存句子中当前正在比较字段的首地址。BX、SI、DI 三个寄存器的作用如图 2.7 所示。

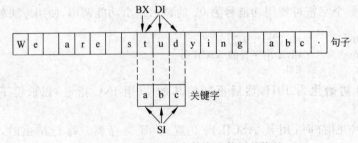

图 2.7　在查找匹配字符串中使用的指针

(3) 输出信息。用功能调用 09h 分"找到"或"未找到"两种情况分别显示不同的信息。在"找到"时，还要求显示出匹配字符串在句子中的位置。我们知道，在"找到"时，BX

寄存器的内容为匹配字符串的首地址,将此值减去句子的首地址,再将差值加1即是所要的匹配字符串在句子中的位置。可将位置值转换为十六进制数从屏幕上显示出来。这一部分程序可参考清华大学出版社1991年出版的《IBM-PC汇编语言程序设计》一书中例6.3 decihex程序中的binihex子过程。

2.2 分支程序设计

一、示例

例2.3 统计学生成绩result

设有10个学生的成绩分别为56、69、84、82、73、88、99、63、100和80分。试编制程序分别统计低于60分、60~69分、70~79分、80~89分、90~99分及100分的人数,并存放到s5、s6、s7、s8、s9及s10单元中。

这一题目的算法很简单,成绩分等部分采用分支结构,统计所有成绩则用循环结构完成。程序框图如图2.8所示。图2.9为程序清单。图2.10表示程序的运行情况。

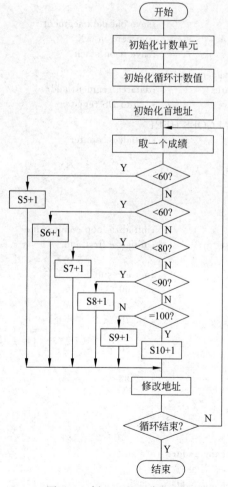

图2.8 例2.3的程序框图

```
;PROGRAM TITLE GOES HERE——result
;****************************************
datarea     segment                         ;define data segment
  grade     dw      56,69,84,82,73,88,99,63,100,80
  s5        dw      0
  s6        dw      0
  s7        dw      0
  s8        dw      0
  s9        dw      0
  s10       dw      0
datarea ends
;****************************************
prognam segment                             ;define code segment
;----------------------------------------
main        proc    far                     ;main part of program
            assume cs:prognam,ds:datarea
start:                                      ;starting execution address
;set up stack for return
            push    ds                      ;save old data segment
            sub     ax,ax                   ;put zero in AX
            push    ax                      ;save it on stack
;set DS register to current data segment
            mov     ax,datarea              ;datarea segment addr
            mov     ds,ax                   ;  into DS register
;MAIN PART OF PROGRAM GOES HERE
            mov     s5,0                    ;initialize counter
            mov     s6,0
            mov     s7,0
            mov     s8,0
            mov     s9,0
            mov     s10,0
            mov     cx,10                   ;initialize loop count value
            mov     bx,offset grade         ;initialize first addr
compare:
            mov     ax,[bx]                 ;get a result
            cmp     ax,60                   ;＜60 ?
            jl      five
            cmp     ax,70                   ;＜70 ?
            jl      six
            cmp     ax,80                   ;＜80 ?
            jl      seven
            cmp     ax,90                   ;＜90 ?
            jl      eight
            cmp     ax,100                  ;＝100 ?
            jne     nine
            inc     s10
            jmp     short change-addr
nine:       inc     s9
            jmp     short change-addr
eight:      inc     s8
```

```
          jmp         short change_addr
seven:    inc         s7
          jmp         short change_addr
six:      inc         s6
          jmp         short change_addr
five:     inc         s5
change_addr:
          add         bx,2
          loop        compare
          ret                                   ;return to DOS
main      endp                                  ;end of main part of program
;————————————————————————————————————
prognam ends
;* * * * * * * * * * * * * * * * * * * * * * * * * * * * * * *
          end         start                     ;end assembly
```

图 2.9 例 2.3 的程序清单

```
C:\ >debug result.exe
—u
19F2:0000  1E              PUSH    DS
19F2:0001  2BC0            SUB     AX,AX
19F2:0003  50              PUSH    AX
19F2:0004  B8F019          MOV     AX,19F0
19F2:0007  8ED8            MOV     DS,AX
19F2:0009  C70614000000    MOV     WORD PTR [0014],0000
19F2:000F  C70616000000    MOV     WORD PTR [0016],0000
19F2:0015  C70618000000    MOV     WORD PTR [0018],0000
19F2:001B  C7061A000000    MOV     WORD PTR [001A],0000
—u
19F2:0021  C7061C000000    MOV     WORD PTR [001C],0000
19F2:0027  C7061E000000    MOV     WORD PTR [001E],0000
19F2:002D  B90A00          MOV     CX,000A
19F2:0030  BB0000          MOV     BX,0000
19F2:0033  8B07            MOV     AX,[BX]
19F2:0035  3D3C00          CMP     AX,003C
19F2:0038  7C32            JL      006C
19F2:003A  3D4600          CMP     AX,0046
19F2:003D  7C27            JL      0066
19F2:003F  3D5000          CMP     AX,0050
—u
19F2:0042  7C1C            JL      0060
19F2:0044  3D5A00          CMP     AX,005A
19F2:0047  7C11            JL      005A
19F2:0049  3D6400          CMP     AX,0064
19F2:004C  7506            JNZ     0054
19F2:004E  FF061E00        INC     WORD PTR [001E]
19F2:0052  EB1C            JMP     0070
19F2:0054  FF061C00        INC     WORD PTR [001C]
19F2:0058  EB16            JMP     0070
19F2:005A  FF061A00        INC     WORD PTR [001A]
19F2:005E  EB10            JMP     0070
```

```
19F2:0060   FF061800      INC        WORD PTR [0018]
-u
19F2:0064   EB0A          JMP        0070
19F2:0066   FF061600      INC        WORD PTR [0016]
19F2:006A   EB04          JMP        0070
19F2:006C   FF061400      INC        WORD PTR [0014]
19F2:0070   83C302        ADD        BX,+02
19F2:0073   E2BE          LOOP       0033
19F2:0075   CB            RETF
19F2:0076   C2D0D8        RET        D8D0
19F2:0079   7319          JNB        0094
19F2:007B   8A46DC        MOV        AL,[BP-24]
19F2:007E   B400          MOV        AH,00
19F2:0080   89C6          MOV        SI,AX
19F2:0082   884ADE        MOV        [BP+SI-22],CL
-g09
AX=19F0  BX=0000  CX=0096  DX=0000  SP=FFFC  BP=0000  SI=0000  DI=0000
DS=19F0  ES=19E0  SS=19F0  CS=19F2  IP=0009   NV UP DI PL ZR NA PE NC
19F2:0009  C70614000000   MOV        WORD PTR [0014],0000        DS:0014=0000
-d0
19F0:0000  38 00 45 00 54 00 52 00-49 00 58 00 63 00 3F 00   8.E.T.R.I.X.c.?.
19F0:0010  64 00 50 00 00 00 00 00-00 00 00 00 00 00 00 00   d.P.............
19F0:0020  1E 2B C0 50 B8 F0 19 8E-D8 C7 06 14 00 00 00 C7   .+@P8p..XG.....G
19F0:0030  06 16 00 00 00 C7 06 18-00 00 00 C7 06 1A 00 00   .....G.....G....
19F0:0040  00 C7 06 1C 00 00 00 C7-06 1E 00 00 00 B9 0A 00   .G.....G......9.
19F0:0050  BB 00 00 8B 07 3D 3C 00-7C 32 3D 46 00 7C 27 3D   ;....=<.|2=F.|'=
19F0:0060  50 00 7C 1C 3D 5A 00 7C-11 3D 64 00 75 06 FF 06   P.|.=Z.|.=d.u...
19F0:0070  1E 00 EB 1C FF 06 1C 00-EB 16 FF 06 1A 00 EB 10   ..k.....k.....k.
-g75
AX=0050  BX=0014  CX=0000  DX=0000  SP=FFFC  BP=0000  SI=0000  DI=0000
DS=19F0  ES=19E0  SS=19F0  CS=19F2  IP=0075   NV UP DI PL NZ NA PE NC
19F2:0075  CB            RETF
-d0
19F0:0000  38 00 45 00 54 00 52 00-49 00 58 00 63 00 3F 00   8.E.T.R.I.X.c.?.
19F0:0010  64 00 50 00 01 00 02 00-01 00 04 00 01 00 01 00   d.P.............
19F0:0020  1E 2B C0 50 B8 F0 19 8E-D8 C7 06 14 00 00 00 C7   .+@P8p..XG.....G
19F0:0030  06 16 00 00 00 C7 06 18-00 00 00 C7 06 1A 00 00   .....G.....G....
19F0:0040  00 C7 06 1C 00 00 00 C7-06 1E 00 00 00 B9 0A 00   .G.....G......9..
19F0:0050  BB 00 00 8B 07 3D 3C 00-7C 32 3D 46 00 7C 27 3D   ;....=<.|2=F.|'=
19F0:0060  50 00 7C 1C 3D 5A 00 7C-11 3D 64 00 75 06 FF 06   P.|.=Z.|.=d.u...
19F0:0070  1E 00 EB 1C FF 06 1C 00-EB 16 FF 06 1A 00 EB 10   ..k.....k.....k.
-q
```

图 2.10 例 2.3 的运行情况

例 2.4 显示月份名 direct

试编写一程序,要求根据用户键入的月份数在终端上显示该月的英文缩写名。

在《IBM-PC 汇编语言程序设计》一书的 5.2.3 节中,介绍了用跳跃表法使程序能根据不同的条件转移到多个程序分支去。在这种方法里,对地址表的访问是根据给定的条件直接计算出相应的表格地址而取得其内容的。这一思想可用在表格查找中。本例要求

根据月份数给出月份名,我们可以建立一个月份表:

```
MONTAB      DB      'JAN'
            DB      'FEB'
            DB      'MAR'
                    ：
```

可以看出,'JAN'的地址为 MONTAB+0,'FEB'的地址为 MONTAB+3,'MAR'的地址为 MONTAB+6,……,根据用户给定的月份数可以计算出与其对应的表格地址,计算方法是

$$MONTAB+(月份数-1)*3$$

应该注意,用户键入的月份数是 ASCII 码,必须转换为数字才能用上述公式进行计算。

根据上述算法,可画出程序框图如图 2.11 所示。程序清单如图 2.12 所示。图 2.13 为该程序的运行情况。

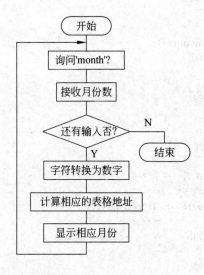

图 2.11　例 2.4 的程序框图

```
;PROGRAM TITLE GOES HERE--DIRECT
;direct(com) Direct table access
;************************************************
datasg      segment     para        'data'
three       db          3
mess        db          'month? ',13,10,'$'
monin       label       byte
  max       db          3
  act       db          ?
  mon       db          3 dup(?)
;
alfmon      db          '??? ',13,10,'$'
montab      db          'JAN','FEB','MAR','APR','MAY','JUN'
            db          'JUL','AUG','SEP','OCT','NOV','DEC'
;
datasg      ends
;************************************************
```

```
codesg   segment   para    'code'
         assume cs:codesg,ds:datasg,es:datasg
main     proc      far
         push      ds
         sub       ax,ax
         push      ax
;
         mov       ax,datasg
         mov       ds,ax
         mov       es,ax
;        Input month:
;        ------
start:
         lea       dx,mess
         mov       ah,09
         int       21h
         lea       dx,monin
         mov       ah,0ah
         int       21h
         mov       dl,13
         mov       ah,02
         int       21h
         mov       dl,10
         mov       2h,02
         int       21h
         cmp       act,0
         je        exit
;        Convert ASCII to binary:
;        -----------
         mov       ah,30h              ;Set up month
         cmp       act,2
         je        two
         mov       al,mon
         jmp       conv
two:
         mov       al,mon+1
         mov       ah,mon
conv:
         xor       ax,3030h            ;Clear ASCII 3's
         cmp       ah,0                ;Month 01-09?
         jz        loc                 ;   yes--bypass
         sub       ah,ah               ;   no--clear AH
         add       al,10               ;   correct for binary
;        Locate month in table:
;        ---------
loc:
         lea       si,montab
         dec       al                  ;Correct for table
         mul       three               ;Mult AL by 3
         add       si,ax
         mov       cx,03               ;Init'ze 3-char move
         cld
         lea       di,alfmon
```

```
            rep       movsb           ;Move 3 chars
;                     Display alpha month:
;                     ——————————
            lea       dx,alfmon
            mov       ah,09
            int       21h
            jmp       start
;
exit:       ret
main        endp
;——————————————————————————————————————————
codesg      ends
;******************************************
            end       main
```

图 2.12 例 2.4 的程序清单

```
C:\>direct
month?
5
MAY
month?
12
DEC
month?
9
SEP
month?
1
JAN
month?
11
NOV
month?
```

图 2.13 例 2.4 的运行情况

例 2.5 显示错误信息 show-err

在调用 DOS 的文件管理功能时，如果出现错误，DOS 处理程序将把进位标志位置 1，并在 AX 中装入错误码，然后调用 show-err 程序将错误信息显示出来。错误码 1～35、80～83 各表示一种 DOS 错误，36～79 为保留代码，其他代码(小于 1 或大于 83)为无效代码。试编写 show-err 程序，要求根据 AX 中的错误码分别显示错误信息、保留代码信息或无效代码信息。

本例要求按 AX 的内容将错误码分为四类：有效代码中的 1～35 和 80～83，保留代码及无效代码，这可以用分支结构来完成。此外，两类有效代码中分别有 35 种和 4 种不同的信息需要输出，这些信息的长度是不同的(见图 2.15 程序清单数据区中的 ER1～ER35 及 ER80～ER83)，为根据 AX 中的内容取得相应的出错信息，我们建立了 ERTAB1 和 ERTAB2 两个地址表，用跳跃表法可以方便地达到目的。它们的地址计算方法分别是：

$$\text{ERTAB1}+[(\text{AX})-1]*2$$

和

$$\text{ERTAB2}+[(\text{AX})\wedge 11\text{B}]*2$$

其中∧11B(二进制的11)的目的是屏蔽高位的8,以取得ERTAB2表中的地址。

本例的程序框图如图2.14所示,图2.15为程序清单,图2.16为程序的运行情况。为调试本程序,必须进入debug以模拟DOS处理程序中把错误码置入AX的功能。由于debug把输入数字确认为十六进制数,因此输入AX的值均为十六进制数,因而图2.16的显示输出都是正确的,且包括了四类错误码的情况。另外,在每次运行完后都要把IP寄存器重新置为0,以便程序重新运行。

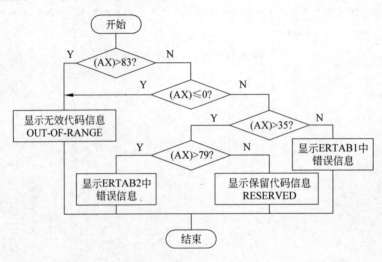

图 2.14 例 2.5 的程序框图

```
TITLE SHOW_ERR--Display DOS function call error messages
;Display a message based on an error code in AX
;All registers are preserved
;* * * * * * * * * * * * * * * * * * * * * * * * * * * * * * * * *
DSEG        SEGMENT   PARA 'DATA'
   CR       EQU       13
   LF       EQU       10
   EOM      EQU       '$'
;
OUT_OF_RANGE DB 'Error code is not in valid range(1-83)'
             DB CR,LF,EOM
RESERVED     DB 'Error code is reserved(36-79)',CR,LF,EOM
ER1          DB 'Invalid function number',CR,LF,EOM
ER2          DB 'File not found',CR,LF,EOM
ER3          DB 'Path not found',CR,LF,EOM
ER4          DB 'Too many open files',CR,LF,EOM
ER5          DB 'Access denied',CR,LF,EOM
ER6          DB 'Invalid handle',CR,LF,EOM
ER7          DB 'Memory control blocks destroyed',CR,LF,EOM
ER8          DB 'Insufficient memory',CR,LF,EOM
```

```
ER9         DB 'Invalid memory block address',CR,LF,EOM
ER10        DB 'Invalid environment',CR,LF,EOM
ER11        DB 'Invalid format',CR,LF,EOM
ER12        DB 'Invalid access code',CR,LF,EOM
ER13        DB 'Invalid data',CR,LF,EOM
ER14        DB 'No such message',CR,LF,EOM
ER15        DB 'Invalid drive was specified',CR,LF,EOM
ER16        DB 'Attempted to remove the current directory'
            DB CR,LF,EOM
ER17        DB 'Not same device',CR,LF,EOM
ER18        DB 'No more files',CR,LF,EOM
ER19        DB 'Disk is write-protected',CR,LF,EOM
ER20        DB 'Unknown unit',CR,LF,EOM
ER21        DB 'Drive not ready',CR,LF,EOM
ER22        DB 'Unknown command',CR,LF,EOM
ER23        DB 'Data error (CRC)',CR,LF,EOM
ER24        DB 'Bad request structure length',CR,LF,EOM
ER25        DB 'Seek error',CR,LF,EOM
ER26        DB 'Unknown media type',CR,LF,EOM
ER27        DB 'Sector not found',CR,LF,EOM
ER28        DB 'Printer out of paper',CR,LF,EOM
ER29        DB 'Write fault',CR,LF,EOM
ER30        DB 'Read fault',CR,LF,EOM
ER31        DB 'General failure',CR,LF,EOM
ER32        DB 'Sharing violation',CR,LF,EOM
ER33        DB 'Lock violation',CR,LF,EOM
ER34        DB 'Invalid disk change',CR,LF,EOM
ER35        DB 'FCB unavailable',CR,LF,EOM
ER80        DB 'File exists',CR,LF,EOM
ER81        DB 'Reserved',CR,LF,EOM
ER82        DB 'Cannot make',CR,LF,EOM
ER83        DB 'Fail on INT 24',CR,LF,EOM
ERTAB1      DW ER1,ER2,ER3,ER4,ER5,ER6,ER7,ER8,ER9,ER10
            DW ER11,ER12,ER13,ER14,ER15,ER16,ER17,ER18
            DW ER19,ER20,ER21,ER22,ER23,ER24,ER25,ER26
            DW ER27,ER28,ER29,ER30,ER31,ER32,ER33,ER34
            DW ER35
ERTAB2      DW ER80,ER81,ER82,ER83
DSEG        ENDS
;*****************************************************
CSEG        SEGMENT PARA 'CODE'
            ASSUME CS : CSEG, DS : DSEG
SHOW_ERR PROC FAR
            PUSH    DS
            SUB     BX,BX
            PUSH    BX
;
            MOV     SI,DSEG             ;initialize DS
            MOV     DS,SI
;
            PUSH    AX                  ;save input error number
```

```
                CMP      AX,83              ;check for error code in range
                JG       O_O_R
                CMP      AX,0
                JG       IN_RANGE
O_O_R:          LEA      DX,OUT_OF_RANGE
                JMP      SHORT DISP_MSG
;Error code is valid,determine with table to use
IN_RANGE:
                CMP      AX,35              ;error code 1 —— 35 ?
                JG       TRY79
                LEA      BX,ERTAB1          ;yes,point to ERTAB1
                DEC      AX
                JMP      FORM_ADDR
TRY79:
                CMP      AX,79              ;error code 36 —— 79 ?
                JG       LAST_4
                LEA      DX,RESERVED        ;yes,display message
                JMP      DISP_MSG
LAST_4:
                LEA      BX,ERTAB2          ;error code 80 —— 83
                AND      AX,3
FORM_ADDR:
                SHL      AX,1               ;point to correct offset
                ADD      BX,AX
                MOV      DX,[BX]            ;put message addr into DX
DISP_MSG:
                MOV      AH,9               ;display messagge string
                INT      21H
                POP      AX
                RET                         ;return to calling program
SHOW_ERR ENDP
;
CSEG            ENDS
;************************************************
                END      SHOW_ERR
```

图 2.15 例 2.5 的程序清单

```
C:\ >debug show_err.exe
-u
19F0:0000  1E          PUSH     DS
19F0:0001  2BDB        SUB      BX,BX
19F0:0003  53          PUSH     BX
19F0:0004  BEF519      MOV      SI,19F5
19F0:0007  8EDE        MOV      DS,SI
19F0:0009  50          PUSH     AX
19F0:000A  3D5300      CMP      AX,0053
19F0:000D  7F05        JG       0014
19F0:000F  3D0000      CMP      AX,0000
19F0:0012  7F06        JG       001A
19F0:0014  8D160000    LEA      DX,[0000]
```

```
19F0:0018   EB26        JMP     0040
19F0:001A   3D2300      CMP     AX,0023
19F0:001D   7F08        JG      0027
19F0:001F   8D1E5A03    LEA     BX,[035A]
-u
19F0:0023   48          DEC     AX
19F0:0024   EB14        JMP     003A
19F0:0026   90          NOP
19F0:0027   3D4F00      CMP     AX,004F
19F0:002A   7F07        JG      0033
19F0:002C   8D162900    LEA     DX,[0029]
19F0:0030   EB0E        JMP     0040
19F0:0032   90          NOP
19F0:0033   8D1EA003    LEA     BX,[03A0]
19F0:0037   250300      AND     AX,0003
19F0:003A   D1E0        SHL     AX,1
19F0:003C   03D8        ADD     BX,AX
19F0:003E   8B17        MOV     DX,[BX]
19F0:0040   B409        MOV     AH,09
19F0:0042   CD21        INT     21
-u
19F0:0044   58          POP     AX
19F0:0045   CB          RETF
19F0:0046   0000        ADD     [BX+SI],AL
19F0:0048   0000        ADD     [BX+SI],AL
19F0:004A   0000        ADD     [BX+SI],AL
19F0:004C   0000        ADD     [BX+SI],AL
19F0:004E   0000        ADD     [BX+SI],AL
19F0:0050   45          INC     BP
19F0:0051   7272        JB      00C5
19F0:0053   6F          DB      6F
19F0:0054   7220        JB      0076
19F0:0056   63          DB      63
19F0:0057   6F          DB      6F
19F0:0058   64          DB      64
19F0:0059   65          DB      65
19F0:005A   206973      AND     [BX+DI+73],CH
19F0:005D   206E6F      AND     [BP+6F],CH
19F0:0060   7420        JZ      0082
19F0:0062   69          DB      69
19F0:0063   6E          DB      6E
-g9
AX=0000   BX=0000   CX=03F8   DX=0000   SP=FFFC   BP=0000   SI=19F5   DI=0000
DS=19F5   ES=19E0   SS=19F0   CS=19F0   IP=0009   NV UP DI PL ZR NA PE NC
19F0:0009 50              PUSH    AX
-rax
AX 0000
:1
-g45
Invalid function number

AX=0001   BX=035A   CX=03F8   DX=0049   SP=FFFC   BP=0000   SI=19F5   DI=0000
DS=19F5   ES=19E0   SS=19F0   CS=19F0   IP=0045   NV UP DI PL NZ NA PE NC
```

```
19F0:0045 CB              RETF
-rip
IP 0045
:0
-rax
AX 0001
:0
-g45
Error code is not in valid range(1-83)

AX=0000  BX=0000  CX=03F8  DX=0000  SP=FFF8  BP=0000  SI=19F5  DI=0000
DS=19F5  ES=19E0  SS=19F0  CS=19F0  IP=0045  NV UP DI PL ZR NA PE NC
19F0:0045 CB              RETF
-rip
IP 0045
:0
-rax
AX 0000
:8
-g45
Insufficient memory

AX=0008  BX=0368  CX=03F8  DX=00DE  SP=FFF4  BP=0000  SI=19F5  DI=0000
DS=19F5  ES=19E0  SS=19F0  CS=19F0  IP=0045  NV UP DI PL NZ AC PO NC
19F0:0045 CB              RETF
-rip
IP 0045
:0
-rax
AX 0008
:14
-g45
Unknown unit

AX=0014  BX=0380  CX=03F8  DX=01F7  SP=FFF0  BP=0000  SI=19F5  DI=0000
DS=19F5  ES=19E0  SS=19F0  CS=19F0  IP=0045  NV UP DI PL NZ AC PO NC
19F0:0045 CB              RETF
-rip
IP 0045
:0
-rax
AX 0014
:23
-g45
FCB unavailable

AX=0023  BX=039E  CX=03F8  DX=0310  SP=FFEX  BP=0000  SI=19F5  DI=0000
DS=19F5  ES=19E0  SS=19F0  CS=19F0  IP=0045  NV UP DI PL NZ NA PO NC
19F0:0045 CB              RETF
-rip
IP 0045
:0
-rax
AX 0023
:24
```

—g45
Error code is reserved (36-79)

AX=0024 BX=0000 CX=03F8 DX=0029 SP=FFE8 BP=0000 SI=19F5 DI=0000
DS=19F5 ES=19E0 SS=19F0 CS=19F0 IP=0045   NV UP DI NG NZ AC PO CY
19F0:0045 CB   RETF
—rip
IP 0045
:0
—rax
AX 0024
:4f
—g45
Error code is reserved (36-79)

AX=004F BX=0000 CX=03F8 DX=0029 SP=FFE4 BP=0000 SI=19F5 DI=0000
DS=19F5 ES=19E0 SS=19F0 CS=19F0 IP=0045   NV UP DI PL ZR NA PE NC
19F0:0045 CB   RETF
—rip
IP 0045
:0
—rax
AX 004F
:50
—g45
File exists

AX=0050 BX=03A0 CX=03F8 DX=0322 SP=FFE0 BP=0000 SI=19F5 DI=0000
DS=19F5 ES=19E0 SS=19F0 CS=19F0 IP=0045   NV UP DI PL NZ NA PE NC
19F0:0045 CB   RETF
—rip
IP 0045
:0
—rax
AX 0050
:53
—g45
Fail on INT 24

AX=0053 BX=03A6 CX=03F8 DX=0349 SP=FFDC BP=0000 SI=19F5 DI=0000
DS=19F5 ES=19E0 SS=19F0 CS=19F0 IP=0045   NV UP DI PL NZ NA PE NC
19F0:0045 CB   RETF
—rip
IP 0045
:0
—rax
AX 0053
:54
—g45
Error code is not in valid range (1-83)

AX=0054 BX=0000 CX=03F8 DX=0000 SP=FFD8 BP=0000 SI=19F5 DI=0000
DS=19F5 ES=19E0 SS=19F0 CS=19F0 IP=0045   NV UP DI PL NZ NA PO NC
19F0:0045 CB   RETF
—q

图 2.16 例 2.5 的运行情况

二、实验题

实验 2.3[*]　分类统计字符个数

1. 题目:分类统计字符个数 COUNT_CHAR。

2. 实验要求:程序接收用户键入的一行字符(字符个数不超过 80 个,该字符串用回车符结束),并按字母、数字及其他字符分类计数,然后将结果存入以 letter、digit 和 other 为名的存储单元中。

3. 提示:

(1) 程序可采用 0AH 功能调用把键入字符直接送到缓冲区中,然后再逐个取出分类计数。也可采用 01H 功能调用在接收字符后先分类计数然后再存入缓冲区中。

(2) 程序需进入 debug 运行并查看计数结果。

有关查表及跳跃表的练习将在下一节和第五章中进行。

2.3　子程序设计

一、示例

例 2.6　显示学生名次表 rank

编制一程序,要求接收从键盘输入的一个班的学生成绩,并存放于 50 字的 grade 数组中,其中 grade+i 保存学号为 i+1 的学生的成绩。然后根据 grade 中的学生成绩,把学生名次填入 50 字的 rank 数组中,其中 rank+i 的内容是学号为 i+1 学生的名次。再按学号顺序把名次从终端上显示出来。

本题要做的主要工作和例 2.2 是完全相同的,只是增加了由用户键入学生成绩及输出学生名次两部分内容。因此,这三部分内容可以用子程序结构来完成。子程序划分的层次图如图 2.17 所示。可以看出,其中 main 为主控模块,其下一层的三个模块为程序的三大部分。现将各模块说明表示如下:

(1) 模块名:main　为总控模块

输入:从键盘输入一个班的学生成绩。

输出:显示一个班的学生名次。

功能:根据输入的学生成绩,计算并显示学生名次。算法如下:

　　一个学生的名次等于成绩高于该生的学生人数加 1。

(2) 模块名:input

输入:以学号为序从键盘输入一个班的学生成绩。各个成绩之间用逗号,隔开,最后
　　　以回车符结束。

输出:把一个班的学生成绩存入 grade 数组。

功能:接收一个班的学生成绩。

　　调用子模块 decibin 把从键盘输入的一个十进制数转换为二进制数。

　　调用子模块 crlf 完成回车、换行功能。

(3) 模块名:rankp

输入:从 grade 数组取得一个班的学生成绩。

输出:以学号为序计算出该班每个学生的名次存入 rank 数组。

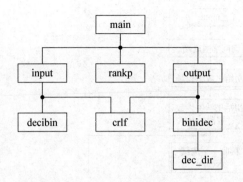

图 2.17 例 2.6 的模块层次图

  功能:计算一个班的学生名次。算法为:一个学生的名次等于成绩高于该生的学生人数加 1。

(4) 模块名:output

输入:从 rank 数组取得一个班的学生名次。

输出:把一个班的学生名次以学号为序在终端上显示出来。

功能:显示一个班的学生名次。

  调用子模块 binidec,以便把 rank 数组中的二进制数转换为十进制数并在终端上显示出来。

  调用子模块 crlf 完成回车、换行功能。

(5) 模块名:decibin

输入:从键盘取得一个十进制数。

输出:把该数转换为二进制数并存入 BX 寄存器中。

功能:把从键盘取得的一个十进制数转换为二进制数,并将该数存入 BX 寄存器中。

(6) 模块名:crlf

输出:向终端发出回车、换行符。

功能:完成一次回车、换行操作。

(7) 模块名:binidec

输入:从 BX 寄存器取得一个二进制数。

输出:在终端屏幕上显示一个十进制数。

功能:把 BX 寄存器中的二进制数转换为十进制数,并在终端屏幕上显示出来。

  调用子模块 dec_div 用来作除法运算并显示字符。

(8) 模块名:dec_div

输入:从 BX 寄存器中取得需转换为十进制的数。

输出:在屏幕上显示一位十进制数。

功能:把 BX 寄存器中的二进制数除以相应的十的幂,并在屏幕上显示一位商。余数保存在 BX 寄存器中。

  有了以上的层次图及模块说明,对程序的全貌就有了基本了解。在图 2.18 中,我们给出了除 rankp 以外的其余各子程序的程序框图。rankp 的框图与图 2.4 相同。图 2.19 是

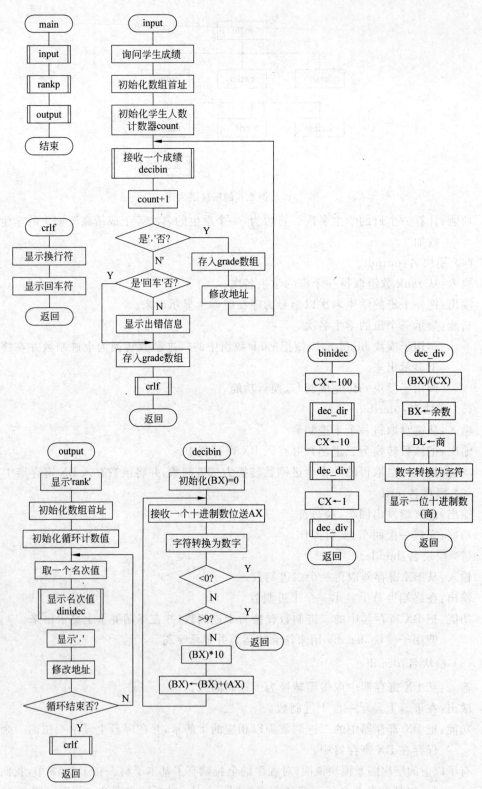

图 2.18 例 2.6 的程序框图

本例的程序清单。图 2.20 是本例的运行情况。
```
;PROGRAM TITLE GOES HERE--Rank
;* * * * * * * * * * * * * * * * * * * * * * * * * * * * * * * * * * * *
datarea    segment                    ;define data segment
  grade            dw      50 dup(?)
  rank             dw      50 dup(?)
  count            dw      ?
  mess1            db      'Grade? $'
  mess2            db      13,10,'Input Error! ',13,10,'$'
  mess3            db      'Rank: $'
datarea    ends
;* * * * * * * * * * * * * * * * * * * * * * * * * * * * * * * * * * * *
prognam segment                       ;define code segment
;------------------------------------------------------------
main       proc    far                ;main part of program
           assume cs:prognam,ds:datarea
start:                                 ;starting execution address
;set up stack for return
           push    ds                 ;save old data segment
           sub     ax,ax              ;put zero in AX
           push    ax                 ;save it on stack
;set DS register to current data segment
           mov     ax,datarea         ;datarea segment addr
           mov     ds,ax              ;  into DS register
;MAIN PART OF PROGRAM GOES HERE
           call    input
           call    rankp
           call    output
           ret
main       endp
;------------------------------------------------------------
input      proc    near
           lea     dx,mess1
           mov     ah,09
           int     21h
;
           mov     si,0
           mov     count,0
enter:
           call    decibin
           inc     count
           cmp     dl,','
           je      store
           cmp     dl,13              ;is it 'return'?
           je      exit2
           jne     error
store:
           mov     grade[si],bx       ;enter the results
           add     si,2               ;  of students
           jmp     enter
```
· 37 ·

```
error:
        lea     dx,mess2
        mov     ah,09
        int     21h
exit2:
        mov     grade[si],bx
        call    crlf
        ret
input   endp
;------------------------------------------------
rankp   proc    near
        mov     di,count
        mov     bx,0
loop1:
        mov     ax,grade[bx]
        mov     word ptr rank[bx],0
        mov     cx,count
        lea     si,grade
next:
        cmp     ax,[si]
        jg      no-count
        inc     word ptr rank[bx]
no-count:
        add     si,2
        loop    next
        add     bx,2
        dec     di
        jne     loop1
        ret                             ;return to DOS
rankp   endp
;------------------------------------------------
output  proc    near
        lea     dx,mess3
        mov     ah,09
        int     21h
;
        mov     si,0
        mov     di,count
next1:
        mov     bx,rank[si]
        call    binidec                 ;display the rank
        mov     dl,','                  ; of students
        mov     ah,02
        int     21h
        add     si,2
        dec     di
        jnz     next1
        call    crlf
        ret
output  endp
;------------------------------------------------
decibin proc    near
```

```
;procedure to convert decimal on keybd to binary.
;result is left in BX register.
         mov      bx,0              ;clear BX for number
;get digit from keyboard, convert to binary
newchar:
         mov      ah,1              ;keyboard input
         int      21h               ;call DOS
         mov      dl,al
         sub      al,30h            ;ASCII to binary
         jl       exit1             ;jump if<0
         cmp      al,9d             ;is it>9d?
         jg       exit1             ;yes,not dec digit
         cbw                        ;byte in AL to word in AX
;(digit is now in AX)
;multiply number in BX by 10 decimal.
         xchg     ax,bx             ;trade digit & number
         mov      cx,10d            ;put 10 dec in CX
         mul      cx                ;number times 10
         xchg     ax,bx             ;trade number & digit
;add digit in AX to number in BX
         add      bx,ax             ;add digit to number
         jmp      newchar           ;get next digit
exit1:   ret                        ;return from decibin
decibin  endp                       ;end of decibin proc
;------------------------------------------------------------
binidec  proc     near
;procedure to convert binary number in BX to decimal
;   on console screen
         push     bx
         push     cx
         push     si
         push     di
         mov      cx,100d           ;divide by 100
         call     dec_div
         mov      cx,10d            ;divide by 10
         call     dec_div
         mov      cx,1d             ;divide by 1
         call     dec_div
         pop      di
         pop      si
         pop      cx
         pop      bx
         ret                        ;return from binidec
binidec  endp
;------------------------------------------------------------
dec_div  proc     near
;sub-subroutine to divide number in BX by number in CX,
;   print quotient on screen
         mov      ax,bx             ;number high half
         mov      dx,0              ;zero out low half
         div      cx                ;divide by CX
         mov      bx,dx             ;remainder into BX
```

```
                mov     dl,al           ;quotient into DL
;print the contents of DL on screen
                add     dl,30h          ;convert to ASCII
                mov     ah,02h          ;display function
                int     21h             ;call DOS
                ret                     ;return from dec_div
dec_div         endp
;----------------------------------------------------------------
crlf            proc    near
;print carriage return and linefeed
                mov     dl,0ah          ;linefeed
                mov     ah,02h          ;display function
                int     21h
;
                mov     dl,0dh          ;carriage return
                mov     ah,02h          ;display function
                int     21h
;
                ret
crlf            endp
;----------------------------------------------------------------
prognam         ends                    ;end of code segment
;* * * * * * * * * * * * * * * * * * * * * * * * * * * * * * * *
                end     start           ;end assembly
```

图 2.19 例 2.6 的程序清单

```
C:\ >rank#
Grade?  67,76,34,100,86,97,54,99,75,98
Rank：  008,006,010,001,005,004,009,002,007,003,

C:\ >rank#
Grade?  78,99,65,89,74,98,84
Rank：  005,001,007,003,006,002,004,

C:\ >rank#
Grade?  87,99,100,76.
Input Error!
Rank：  003,002,001,004,
```

图 2.20 例 2.6 的运行情况

例 2.7 计算工资 scremp

编写一程序,接收用户输入的工作时间及工资率(即每小时的工资数),显示计算而得的工资数。

本程序由三部分组成:输入工作时间和工资率;计算工资;显示工资值。在输入、输出部分,与例 2.6 一样,必须考虑字符与数字的转换以及十化二、二化十问题。除此之外,还应该注意到本例中的输入数可能是小数。在这里并不需要使用浮点数格式来进行计算,只是在计算中必须处理小数。我们采用在接收输入数时记录小数点后的位数,并把两个输入数的小数点后位数之和存放在 nodec 单元中。在计算工资的乘法中,并不考虑小数点的存在,而输出的工资数又只需取小数点后的二位数,为此我们用 shift 单元记录移位

因子,用 adjust 单元记录舍入值。对于 nodec 的不同值可以分以下三种情况处理:

(1) nodec>6

我们知道,对于 16 位整数而言,机器允许的最大数为 65535。对于 nodec>6 的数,移位因子将≥100000,该数已超过机器允许的范围,因此本例限制 nodec 的值必须≤6。如 nodec>6,则作为溢出处理,此时将输出值置为 0。

(2) nodec=3~6

此时,

$$\text{移位因子} \quad shift=(10)^{nodec-2}$$

$$\text{舍入值} \quad adjust=\frac{1}{2}shift$$

例如:输入工作时间为 8.52,工资率为 10.25,则乘积为

$$product=852\times1025=873300$$

移位因子为 $\quad shift=(10)^{nodec-2}=(10)^{4-2}=100$

舍入值为 $\quad adjust=\frac{1}{2}shift=50$

作舍入及移位处理:

$$(product+adjust)/shift=(873300+50)/100=8733$$

又如,输入工作时间为 65.245,工资率为 8.52,则

$$product=65245\times852=55588740$$

$$shift=(10)^{5-2}=1000$$

$$adjust=500$$

作舍入及移位处理:

$$(55588740+500)/1000=55589240/1000$$
$$=55589$$

经处理后的值只是取得了答案的有效值,并未考虑小数点的位置,这个问题将在输出显示时解决。

(3) nodec=0~2

在这种情况下,乘积的结果不必作舍入及移位处理,只需记录 nodec 值,并在输出显示时解决小数点的位置即可。

在计算中还需要说明一个问题:由于采用整数运算,要求输入数不超过 65535,任一输入数超过该值就作溢出处理,在这里我们用输出 0 来表示。此外,两个输入数的乘积可能得到 32 位二进制数的结果,这用一般的字运算乘法指令就可得到。对于这样的双精度数作除法运算时,尽管我们已经限制 shift 的值不超过 65535,但字运算的除法指令要求双精度的被除数和单精度的除数相除,其结果应该是单精度的商,否则就作为溢出处理。为了避免这种溢出情况的发生,我们采用以下程序段来作除法。设在以下程序段运行前,我们已取得双精度被除数在 DX:AX 中,除数在 shift 单元中,除法运算结果的商在 DX:AX 中。

```
        mov     tempdx,dx
        mov     tempax,ax
        mov     ax,dx
```

```
        mov     dx,0
        div     shift
        mov     tempdx,ax
        mov     ax,tempax
        div     shift
        mov     dx,tempdx
        mov     tempax,ax
```

顺便说明一下,在输出的二化十计算中所用除法也采用这种方法进行。

本例的模块层次图如图 2.21 所示。各模块说明如下:

(1) 模块名:begin 为总控模块

输入:接收从键盘输入的工作时间 hour 和工资率 rate。

输出:在屏幕上显示工资值 wage。

功能:根据工作时间和工资率计算工资。

$$wage = hour * rate$$

调用:　　q10scr　　清除屏幕;
　　　　　q20curs　　置光标位置;
　　　　　b10inpt　　接收 hour 和 rate;
　　　　　d10hour　　把 hour 的 ASCII 转换为二进制数;
　　　　　e10rate　　把 rate 的 ASCII 转换为二进制数;
　　　　　f10mult　　计算工资 wage;
　　　　　g10wage　　把 wage 的二进制数转换为 ASCII 码;
　　　　　h10disp　　显示 wage。

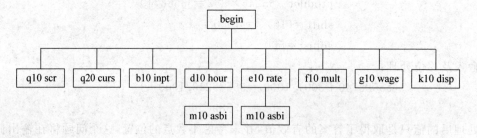

图 2.21　例 2.7 的模块层次图

(2) 模块名:q10scr

功能:清除屏幕。

(3) 模块名:q20curs

功能:置光标位置。

(4) 模块名:b10inpt

输入:接收从键盘输入的以小时为单位的工作时间及工资率(元/小时)。

输出:把工作时间存入 hrspar 缓冲区,把工资率存入 ratepar 缓冲区。

功能:接收从键盘输入的工作时间及工资率,分别存入相应的缓冲区中。

(5) 模块名:d10hour

输入:从 hrspar 中取出工作时间。

输出:把转换为二进制的工作时间存入 binhrs 单元中。

功能:调用子过程 m10asbi,把工作时间从 ASCII 码转换为二进制数。

(6) 模块名:e10rate

输入:从 ratepar 中取出工资率。

输出:把转换为二进制的工资率存入 binrate 单元中。

功能:调用子过程 m10asbi,把工资率从 ASCII 码转换为二进制数。

(7) 模块名:m10asbi

输入:根据调用程序给出的指针(在 SI 中)以及字符个数(在 CX 中)取得一个 ASCII 字符串。

输出:将 ASCII 字符串转换为二进制数,结果存放在 binval 单元中。

功能:把 ASCII 字符串转换为二进制数。同时记录输入数的小数点后的位数,累计于 nodec 单元中。

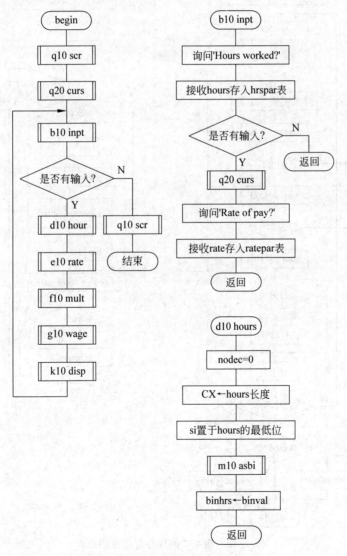

图 2.22 例 2.7 的部分程序框图(1)

使用 decind 作为小数点标志位。如输入数有小数点则 decind 内容为 1,否则为 0。

(8) 模块名:f10mult

输入:从 binhrs 中取得工作时间,从 binrate 中取得工资率。

输出:根据工作时间及工资率计算而得的工资值存放在 dx:ax 中。

功能:把工作时间和工资率的乘积经舍入和移位处理后得到的二进制工资值存放在 dx:ax 中。

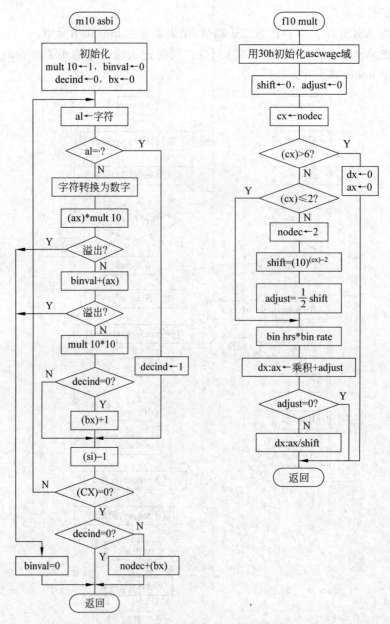

图 2.22 例 2.7 的部分程序框图(2)

(9) 模块名:g10wage

输入:dx:ax 中的二进制工资数以及 nodec 单元中的小数点后的位数(0,1 或 2)。

输出:把二进制的工资数转换为 ASCII 码存放在以 ascwage 为首地址的字符串中。

功能:把 dx:ax 中的二进制工资数转换为十进制数的 ASCII 形式,并插入小数点,存放在以 ascwage 为首地址的字符串中。

(10) 模块名:k10disp

输入:ascwage 中的字符串。

输出:把字符串在屏幕上显示出来。

功能:显示工资数。

以上给出了所有的模块说明。图 2.23 给出了本例中几个主要模块的程序框图。图 2.24 给出本例的全部程序清单。图 2.25 则为本例的运行结果。可以看出,其中工资数为 0 的情况除有一例为 0∗0 以外,其余均为输入数溢出或两个输入数小数点后的位数总和大于 6 的情况。

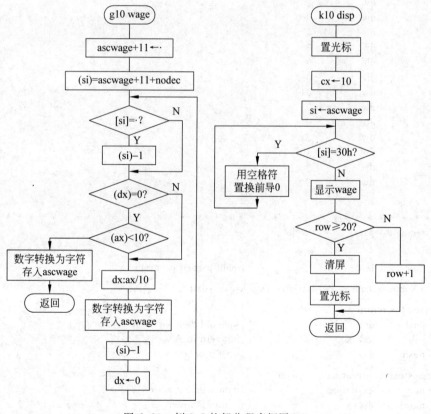

图 2.23 例 2.7 的部分程序框图(3)

```
;PROGRAM TITLE GOES HERE－－SCREMP
; Enter hours & rate, display wage
;∗∗∗∗∗∗∗∗∗∗∗∗∗∗∗∗∗∗∗∗∗∗∗∗∗∗∗∗∗∗∗∗∗∗∗∗∗∗∗∗∗∗∗∗∗∗∗
stacksg    segment    para stack 'stack'
```

```
            dw        32 dup(?)
stacksg     ends
;* * * * * * * * * * * * * * * * * * * * * * * * * * * * * * * * * * *
datasg      segment   para 'data'
   hrspar         label     byte            ;Hours parameter list;
   maxhlen        db        6               ;___ ____ __
   acthlen        db        ?
   hrsfld         db        6 dup(?)

   ratepar        label     byte            ;Rate parameter list;
   maxrlen        db        6               ;____ ___ ___
   actrlen        db        ?
   ratefld        db        6 dup(?)
   messg1         db        'Hours worked? ','$'
   messg2         db        'Rate of pay? ','$'
   messg3         db        'Wage = '
   ascwage        db        14 dup(30h),13,10,'$'

   messg4         db        13,10,'Overflow! ',13,10,'$'
   adjust         dw        ?
   binval         dw        0
   binhrs         dw        0
   binrate        dw        0
   col            db        0
   decind         db        0
   mult10         dw        01
   nodec          dw        0
   row            db        0
   shift          dw        ?
   tenwd          dw        10
   tempdx         dw        ?
   tempax         dw        ?
datasg      ends
;* * * * * * * * * * * * * * * * * * * * * * * * * * * * * * * * * * *
codesg      segment   para      'code'
;_ _ _ _ _ _ _ _ _ _ _ _ _ _ _ _ _ _ _ _ _ _ _ _ _ _ _ _ _ _ _ _ _ _
begin       proc      far                   ;main part of program
            assume cs:codesg,ds:datasg,ss:stacksg,es:datasg
;set up stack for return
            push      ds                    ;save old data segment
            sub       ax,ax                 ;put zero in AX
            push      ax                    ;save it on stack
;set DS register to current data segment
            mov       ax,datasg             ;data segment addr
            mov       ds,ax                 ;   into DS register
            mov       es,ax                 ;   into ES register
;MAIN PART OF PROGRAM GOES HERE
            mov       ax,0600h
            call      q10scr                ;clear screen
            call      q20curs               ;set cursor
a20loop:
```

```
                call    b10inpt             ;Accept hours & rate
                cmp     acthlen,0           ;End of input?
                je      a30
                call    d10hour             ;Convert hours to binary
                call    e10rate             ;Convert rate to binary
                call    f10mult             ;Calc wage, round
                call    g10wage             ;Convert wage to ASCII
                call    k10disp             ;Display wage
                jmp     a20loop
a30:
                mov     ax,0600h
                call    q10scr              ;Clear screen
                ret                         ;return to DOS
begin           endp                        ;end of main part of program
;------------------------------------------------------------------
;                       Input hours & rate:
;                       -------------------
b10inpt         proc    near
                lea     dx,messg1           ;Prompt for hours
                mov     ah,09h
                int     21h
                lea     dx,hrspar           ;Accept hours
                mov     ah,0ah
                int     21h
                cmp     acthlen,0           ;No hours? (indicates end)
                jne     b20
                ret                         ;If so, return to a20loop
b20:
                mov     col,25              ;Set column
                call    q20curs
                lea     dx,messg2           ;Prompt for rate
                mov     ah,09h
                int     21h
                lea     dx,ratepar          ;Accept rate
                mov     ah,0ah
                int     21h
                ret
b10inpt         endp
;------------------------------------------------------------------
;                       Process hours:
;                       --------------
d10hour         proc    near
                mov     nodec,0
                mov     cl,acthlen
                sub     ch,ch
                lea     si,hrsfld-1         ;Set right pos'n
                add     si,cx               ;  of hours
                call    m10asbi             ;Convert to binary
                mov     ax,binval
                mov     binhrs,ax
                ret
d10hour         endp
```

```
;----------------------------------------------------------------
;                       Process rate:
;                       ------------
e10rate  proc    near
         mov     cl,actrlen
         sub     ch,ch
         lea     si,ratefld-1           ;Set right pos'n
         add     si,cx                  ;  of rate
         call    m10asbi                ;Convert to binary
         mov     ax,binval
         mov     binrate,ax
         ret
e10rate  endp
;----------------------------------------------------------------
;                       Multiply, round, & shift:
;                       ------------------------
f10mult  proc    near
         mov     cx,07
         lea     di,ascwage             ;Set ASCII wage
         mov     ax,3030h               ;   to 30's
         cld
         rep     stosw

         mov     shift,10
         mov     adjust,0
         mov     cx,nodec
         cmp     cl,06                  ;If more than 6
         ja      f40                    ;  decimals, error
         dec     cx
         dec     cx
         jle     f30                    ;Bypass if 0, 1, 2 decs
         mov     nodec,02
         mov     ax,01
f20:
         mul     tenwd
         loop    f20                    ;Calculate shift factor

         mov     shift,ax
         shr     ax,1                   ;Calculate round value
         mov     adjust,ax
f30:
         mov     ax,binhrs
         mul     binrate                ;Calculate wage
         add     ax,adjust              ;Round wage
         adc     dx,0
         mov     tempdx,dx              ;Save DX:AX
         mov     tempax,ax
;
         cmp     adjust,0               ;No shift required?
         jz      f50
;
         mov     ax,dx                  ;Shift use double-precision
         mov     dx,0                   ;   DIV,
         div     shift                  ;     quotient into DX:AX
```

```
                mov     tempdx,ax
                mov     ax,tempax
                div     shift
                mov     dx,tempdx
                mov     tempax,ax
                jmp     f50
;
f40:
                mov     ax,0                    ;Overflow
                mov     dx,0
f50:
                ret                             ;Return
f10mult endp
;------------------------------------------------------------
;               Convert to ASCII:
;               ---------------
g10wage proc    near
                lea     si,ascwage+11           ;Set decimal pt.
                mov     byte ptr[si],'.'
                add     si,nodec                ;Set right start pos'n
g30:
                cmp     byte ptr[si],'.'
                jne     g35                     ;Bypass if at dec pos'n
                dec     si
g35:
                cmp     dx,0                    ;If DX:AX < 10,
                jnz     g40
                cmp     ax,0010
                jb      g50                     ;  operation finished
g40:
                mov     ax,dx
                mov     dx,0
                div     tenwd
                mov     tempdx,ax
                mov     ax,tempax
                div     tenwd                   ;Remainder is ASCII digit
                mov     tempax,ax
                or      dl,30h
                mov     [si],dl                 ;Store ASCII character
                dec     si
                mov     dx,tempdx
                jmp     g30
g50:
                or      al,30h                  ;Store last ASCII
                mov     [si],al                 ;  character
                ret
g10wage endp
;------------------------------------------------------------
;               Display wage:
;               ------------
k10disp proc    near
                mov     col,50                  ;Set column
                call    q20curs
```

```
              mov       cx,10
              lea       si,ascwage
k20:                                            ;Clear leading zeros
              cmp       byte ptr[si],30h
              jne       k30                     ;   to blanks
              mov       byte ptr[si],20h
              inc       si
              loop      k20
k30:
              lea       dx,messg3               ;Display
              mov       ah,09
              int       21h
              cmp       row,20                  ;Bottom of screen?
              jae       k80
              inc       row                     ;  no——increment row
              jmp       k90
k80:
              mov       ax,0601h                ;   yes——
              call      q10scr                  ;   scroll &
              mov       col,0                   ;   set cursor
              call      q20curs
k90:          ret
k10disp       endp
;———————————————————————————————————————————————
;             Convert ASCII to binary:
;             ——————————
m10asbi       proc      near
              mov       mult10,01
              mov       binval,0
              mov       decind,0
              sub       bx,bx
m20:
              mov       al,[si]                 ;Get ASCII character
              cmp       al,'.'                  ;Bypass if dec pt.
              jne       m40
              mov       decind,01
              jmp       m90
m40:
              and       ax,000fh
              mul       mult10                  ;Multiply by factor
              jc        overflow
              add       binval,ax               ;Add to binary
              jc        overflow
              mov       ax,mult10               ;Calculate next
              mul       tenwd                   ;  factor * 10
              mov       mult10,ax
              cmp       decind,0                ;Reached decimal pt?
              jnz       m90
              inc       bx                      ;   yes——add to count
m90:
              dec       si
              loop      m20

              cmp       decind,0                ;End of loop
              jz        m100                    ;And decimal pt?
              add       nodec,bx                ;   yes——add to total
```

```
                jmp       m100
overflow:
                mov       binval,0
m100:           ret
m10asbi endp
;------------------------------------------------------------
;                Scroll screen:
;                -------------
q10scr   proc   near                      ;AX set on entry
         mov    bh,07                     ;Color (07 for BW)
         sub    cx,cx
         mov    dx,184fh
         int    10h
         ret
q10scr   endp
;------------------------------------------------------------
;                Set cursor:
;                ----------
q20curs  proc   near
         mov    ah,2
         sub    bh,bh
         mov    dh,row
         mov    dl,col
         int    10h
         ret
q20curs  endp
;------------------------------------------------------------
codesg   ends                             ;end of code segment
;************************************************************
         end    begin                     ;end assembly
```

图 2.24 例 2.7 的程序清单

Hours worked? 65535	Rate of pay? 65535	Wage= 4294836225.00
Hours worked? 0	Rate of pay? 0	Wage= 0.00
Hours worked? 1	Rate of pay? 1	Wage= 1.00
Hours worked? 625	Rate of pay? 700	Wage= 437500.00
Hours worked? 65535	Rate of pay? 999.9	Wage= 65528446.50
Hours worked? 65535	Rate of pay? 99.99	Wage= 6552844.65
Hours worked? 65535	Rate of pay? 9.999	Wage= 655284.47
Hours worked? 65535	Rate of pay? .9999	Wage= 65528.45
Hours worked? 999.9	Rate of pay? .9999	Wage= 999.80
Hours worked? 99.99	Rate of pay? .9999	Wage= 99.98
Hours worked? 9.999	Rate of pay? .9999	Wage= 0.00
Hours worked? 6553	Rate of pay? 999.9	Wage= 6552344.70
Hours worked? 655.3	Rate of pay? 999.9	Wage= 655234.47
Hours worked? 655.3	Rate of pay? 99.99	Wage= 65523.45
Hours worked? 65.53	Rate of pay? 99.99	Wage= 6552.34
Hours worked? 65536	Rate of pay? 98	Wage= 0.00
Hours worked? 12	Rate of pay? 70000	Wage= 0.00
Hours worked? 90000	Rate of pay? 67	Wage= 0.00
Hours worked?		

图 2.25 例 2.7 的运行情况

例 2.8 HANOI 塔题

编写解 HANOI 塔谜题的程序。在这个谜题中，A 轴自下而上地叠有大小逐渐减小的 N 个盘子（见图 2.26），现要求把它们移到 C 轴上并保持原来的次序。移动时允许把盘子暂时存放在 B 轴上，但移动盘子必须遵循以下两条规则：(1)一次只能有一个盘子从一个轴移到另一个轴上；(2)一个盘子在任何时候都不能放在比它小的盘子上面。

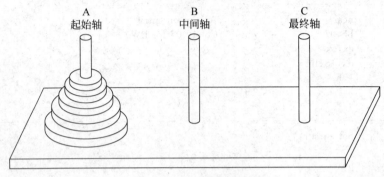

图 2.26 HANOI 塔谜题

这是一个经典的递归子程序的例子。我们用 N 表示盘子数，并从小到大把盘子编号为 1,2,…,N。用 X,Y,Z 表示起始轴、中间轴和最终轴。用 uiv 表示第 i 个盘子从 u 轴移到 v 轴，其中 u,v 可以是 X,Y,Z 中的任一个轴。这样 HANOI 塔题可用递归定义描述如下：

基数：HANOI(1,X,Y,Z)　显示 X1Z

归纳步骤：HANOI(N,X,Y,Z)　(N>1)　做以下三步：

(1) 执行 HANOI(N−1,X,Z,Y)

(2) 显示 XNZ

(3) 执行 HANOI(N−1,Y,X,Z)

根据这一算法可以分析出 HANOI(3,A,B,C) 的执行过程如图 2.27 所示。图 2.28 给出了本例的模块层次图。各模块说明如下：

(1) 模块名：main　为总控模块。

输入：接收从键盘输入的盘子数 N 及起始、中间和最终轴名，存放在 BX、CX、SI 和 DI 寄存器中。

输出：以 uiv 形式，顺序显示出盘子的移动办法。

功能：用递归算法计算并显示出 HANOI 塔的盘子移动办法。算法如下：

　　　　基数：HANOI(1,X,Y,Z) 显示 X1Z

　　　　归纳步骤：HANOI(N,X,Y,Z)　(N>1)

　　　　　　① 执行 HANOI(N−1,X,Z,Y)

　　　　　　② 显示 XNZ

　　　　　　③ 执行 HANOI(N−1,Y,X,Z)

调用子程序 decibin 接收盘子数 N；

调用子程序 crlf 实现回车、换行功能；

调用子程序 hanoi 求解 HANOI 塔谜题。

（2）模块名：decibin

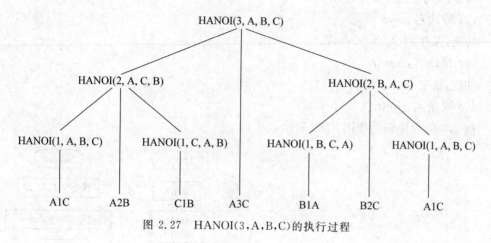

图 2.27　HANOI(3,A,B,C)的执行过程

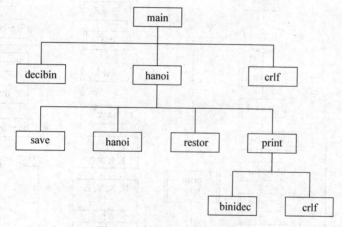

图 2.28　例 2.8 的模块层次图

输入：从键盘接收盘子数 N。

输出：把转换为二进制数的 N 值存入 BX 寄存器中。

功能：把从键盘接收的十进制数转换为二进制数并存入 BX 寄存器中。

（3）模块名：hanoi

输入：从 BX 寄存器中取得盘子数 N。

输出：显示盘子移动办法。

功能：用递归算法计算并显示 HANOI 塔题的结果。

　　递归调用子程序 hanoi；

　　调用子程序 save 保存 N、X、Y、Z；

　　调用子程序 restor 恢复 N、X、Y、Z；

　　调用子程序 print 显示计算结果。

(4) 模块名:crlf

功能:显示回车、换行符。

(5) 模块名:save

功能:保存 N、X、Y、Z 入栈。

(6) 模块名:restor

功能:恢复 N、X、Y、Z。

(7) 模块名:print

例 2.8 的程序框图如图 2.29 所示。

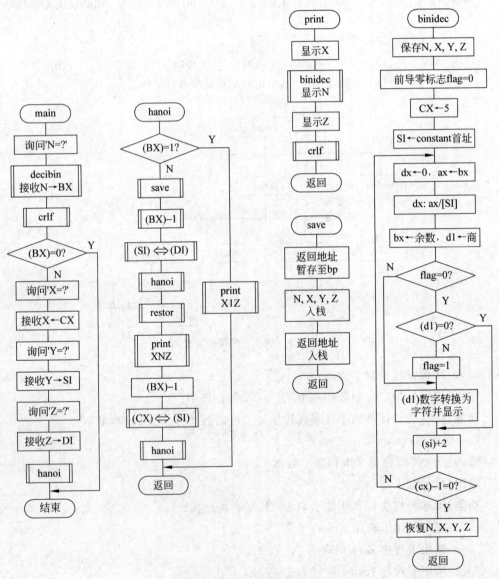

图 2.29　例 2.8 的程序框图

功能:显示 XNZ

调用子程序 binidec,把 N 值从二进制转换为十进制并在屏幕上显示出来。

调用子程序 crlf 以取得回车、换行效果。

(8) 模块名:binidec

输入:从 BX 寄存器中取得 N 值。

输出:把 N 值以十进制形式在屏幕上显示出来。

功能:把 BX 寄存器中的二进制 N 值转换为十进制形式,并在屏幕上显示出来。

本例的各主要模块的程序框图表示于图 2.29。图 2.30 为程序清单。图 2.31 给出了程序在 N=3 和 N=4 时的运行结果,可以看出 N=3 的情况和图 2.27 的分析结果是完全一致的。图 2.32 给出了当 N=3 时的程序运行过程及堆栈的变化,建议读者仔细阅读以弄清递归子程序的调用过程及相应的堆栈状态的变化情况。为使读者阅读方便起见,图 2.33 给出了本例 LST 文件中的 hanoi 子程序部分的 LST 清单,此外还说明以下几点:

① 程序开始运行时用 g59 启动运行程序,0059 为 hanoi 子程序的入口地址,因此程序在接收 N 值及三个轴名后停于断点 0059。此时显示的指令 CMP BX,+01 尚未执行,用 t 命令执行该指令并显示下一条将要执行的指令为 JZ 007C,再用 t 命令执行该指令并显示下一条将要执行的指令为 CALL 008D。008D 为子程序 save 的入口地址,此时,为直接看到子程序执行完后的结果(而不是子程序中的每条指令的执行结果)用 p 命令执行该子程序,程序停于该子程序返回后的第一条指令 DEC BX 处,此时用 d 命令显示的堆栈内容为 save 子程序运行后的堆栈状态。

在图 2.32 所示的运行过程中,对于除 hanoi 以外的所有子程序均使用 p 命令执行,但在执行 hanoi 时,为了解程序的运行细节,使用 t 命令逐条执行。此外,除 print 子程序外,其余子程序执行完后,均用 d 命令显示堆栈的变化。

```
; PROGRAM TITLE GOES HERE－－HANOI
; Solves tower of HANOI puzzle.  Printout sequence of moves
; of N discs from initial spindle X to final spindle Z.
; Using spindle Y for temporery storage.
;*******************************************************
    datarea     segment              ;define data segment
        message1        db      'N=? ',0ah,0dh,'$'
        message2        db      'What is the name of spindle X ? '
                        db      0ah,0dh,'$'
        message3        db      'What is the name of spindle Y ? '
                        db      0ah,0dh,'$'
        message4        db      'What is the name of spindle Z ? '
                        db      0ah,0dh,'$'
        flag            dw      0
        constant        dw      10000,1000,100,10,1
    datarea     ends
;*******************************************************
    prognam     segment              ;define code segment
;-------------------------------------------------------
    main        proc    far
                assume cs:prognam,ds:datarea
```

```
start:
;set up stack for return
        push    ds
        sub     ax,ax
        push    ax
;set DS register to current data segment
        mov     ax,datarea
        mov     ds,ax
;MAIN PART OF PROGRAM GOES HERE
        lea     dx,message1     ; N=?
        mov     ah,09h
        int     21h
        call    decibin         ;read N into BX
        call    crlf
;
        cmp     bx,0            ;if N=0,
        jz      exit            ;  exit
;
        lea     dx,message2     ; X=?
        mov     ah,09h
        int     21h
        mov     ah,01h          ;read X's name into CX
        int     21h
        mov     ah,0
        mov     cx,ax
        call    crlf
;
        lea     dx,message3     ; Y=?
        mov     ah,09h
        int     21h
        mov     ah,01h          ;read Y's name into SI
        int     21h
        mov     ah,0
        mov     si,ax
        call    crlf
;
        lea     dx,message4     ; Z=?
        mov     ah,09h
        int     21h
        mov     ah,01h          ;read Z's name into DI
        int     21h
        mov     ah,0
        mov     di,ax
        call    crlf
;
        call    hanoi           ;call HANOI( N,X,Y,Z )
;
exit:   ret                     ;return to DOS
;
main    endp
;------------------------------------------------------
hanoi   proc    near            ;define subprocedure
```

```
; Solves tower of HANOI puzzle.
; Argement: (BX)=N, (CX)=X, (SI)=Y, (DI)=Z.
            cmp     bx,1
            je      basis               ;if N=1,execute basis
            call    save                ;save N,X,Y,Z
            dec     bx
            xchg    si,di
            call    hanoi               ;call HANOI(N-1,X,Z,Y)
            call    restor              ;restore N,X,Y,Z
            call    print               ;print XNZ
            dec     bx
            xchg    cx,si
            call    hanoi               ;call HANOI(N-1,Y,X,Z)
            jmp     return
basis:      call    print               ;print X1Z
return:     ret                         ;return
hanoi       endp                        ;end subprocedure
;----------------------------------------------------------------
print       proc    near
;print XNZ
            mov     dx,cx               ;print X
            mov     ah,02h
            int     21h
            call    binidec             ;print N
            mov     dx,di               ;print Z
            mov     ah,02h
            int     21h
            call    crlf                ;skip to next line
            ret
print       endp
;----------------------------------------------------------------
save        proc    near
;push N,X,Y,Z onto stack
            pop     bp
            push    bx
            push    cx
            push    si
            push    di
            push    bp
            ret
save        endp
;----------------------------------------------------------------
restor      proc    near
;pop Z,Y,X,N from stack
            pop     bp
            pop     di
            pop     si
            pop     cx
            pop     bx
            push    bp
            ret
restor      endp
```

```
;------------------------------------------------------------
decibin   proc      near
;procedure to convert decimal on keybd to binary.
;result is left in BX register.
          mov       bx,0              ;clear BX for number
;get digit from keyboard, convert to binary
newchar:
          mov       ah,1              ;keyboard input
          int       21h               ;call DOS
          sub       al,30h            ;ASCII to binary
          jl        exit1             ;jump if<0
          cmp       al,9d             ;is it>9d?
          jg        exit1             ;yes, not dec digit
          cbw                         ;byte in AL to word in AX
;(digit is now in AX)
;multiply number in BX by 10 decimal.
          xchg      ax,bx             ;trade digit & number
          mov       cx,10d            ;put 10 dec in CX
          mul       cx                ;number times 10
          xchg      ax,bx             ;trade number & digit
;add digit in AX to number in BX
          add       bx,ax             ;add digit to number
          jmp       newchar           ;get next digit
exit1:    ret                         ;return from decibin
decibin   endp                        ;end of decibin proc
;------------------------------------------------------------
binidec   proc      near
;procedure to convert binary number in BX to decimal
;   on console screen
          push      bx
          push      cx
          push      si
          push      di
          mov       flag,0
          mov       cx,5
          lea       si,constant
;
dec_div:
          mov       ax,bx             ;number high half
          mov       dx,0              ;zero out low half
          div       word ptr[si]      ;divide by contant
          mov       bx,dx             ;remainder into BX
          mov       dl,al             ;quotient into DL
;
          cmp       flag,0            ;have not leading zero
          jnz       print1
          cmp       dl,0
          je        skip
          mov       flag,1
;print the contents of DL on screen
print1:   add       dl,30h            ;convert to ASCII
          mov       ah,02h            ;display function
          int       21h               ;call DOS
```

```
skip:      add      si,2
           loop     dec_div
           pop      di
           pop      si
           pop      cx
           pop      bx
           ret                              ;return from dec_div
binidec    endp
;------------------------------------------------------------
crlf       proc     near
;print carriage return and linefeed
           mov      dl,0ah                  ;linefeed
           mov      ah,02h                  ;display function
           int      21h
;
           mov      dl,0dh                  ;carriage return
           mov      ah,02h                  ;display function
           int      21h
;
           ret
crlf       endp
;------------------------------------------------------------
prognam    ends                             ;end of code segment
;************************************************************
           end      start                   ;end assembly
```

图 2.30 例 2.8 的程序清单

```
C:\ >hanoi#
N=?
3
What is the name of spindle X ?
A
What is the name of spindle Y ?
B
What is the name of spindle Z ?
C
A1C
A2B
C1B
A3C
B1A
B2C
A1C
C:\ >hanoi#
N=?
4
What is the name of spindle X ?
A
What is the name of spindle Y ?
B
What is the name of spindle Z ?
```

C
A1B
A2C
B1C
A3B
C1A
C2B
A1B
A4C
B1C
B2A
C1A
B3C
A1B
A2C
B1C

图 2.31 例 2.8 的运行结果

```
C:\ >debug hanoi#.exe
-g59
N=?
3
What is the name of spindle X ?
A
What is the name of spindie Y ?
B
What is the name of spindie Z ?
C
AX=020D  BX=0003  CX=0041  DX=000D  SP=FFFA  BP=0000  SI=0042  DI=0043
DS=1A04  ES=19F4  SS=1A04  CS=1A0C  IP=0059    NV UP DI PL NZ NA PE NC
1A0C:0059 83FB01        CMP       BX,+01
-d1a04:ffe0
1A04:FFE0  B9 0B 0D 02 03 00 41  00-0D 00 42 00 43 00 00 00    9.....A...B.C..
1A04:FFF0  04 1A 00 00 59 00 0C  1A-D4 16 58 00 00 00 F4 19    ....Y...T.X...t.
-t
AX=020D  BX=0003  CX=0041  DX=000D  SP=FFFA  BP=0000  SI=0042  DI=0043
DS=1A04  ES=19F4  SS=1A04  CS=1A0C  IP=005C    NV UP DI PL NZ NA PO NC
1A0C:005C 7418          JZ        0076
-t
AX=020D  BX=0003  CX=0041  DX=000D  SP=FFFA  BP=0000  SI=0042  DI=0043
DS=1A04  ES=19F4  SS=1A04  CS=1A0C  IP=005E    NV UP DI PL NZ NA PO NC
1A0C:005E E82C00            CALL      008D
-p
AX=020D  BX=0003  CX=0041  DX=000D  SP=FFF2  BP=0061  SI=0042  DI=0043
DS=1A04  ES=19F4  SS=1A04  CS=1A0C  IP=0061    NV UP DI PL NZ NA PO NC
1A0C:0061 4B                DEC       BX
-d1a04:ffe0
1A04:FFE0  B9 0B 0D 02 03 00 41  00-0D 00 61 00 61 00 0C 1A    9.....A...a.a...
1A04:FFF0  D4 16 43 00 42 00 41  00-03 00 58 00 00 00 F4 19    T.C.B.A...X...t.
```

-t

AX=020D BX=0002 CX=0041 DX=000D SP=FFF2 BP=0061 SI=0042 DI=0043
DS=1A04 ES=19F4 SS=1A04 CS=1A0C IP=0062 NV UP DI PL NZ NA PO NC
1A0C:0062 87F7   XCHG   DI,SI
-t

AX=020D BX=0002 CX=0041 DX=000D SP=FFF2 BP=0061 SI=0043 DI=0042
DS=1A04 ES=19F4 SS=1A04 CS=1A0C IP=0064 NV UP DI PL NZ NA PO NC
1A0C:0064 E8F2FF   CALL   0059
-t

AX=020D BX=0002 CX=0041 DX=000D SP=FFF0 BP=0061 SI=0043 DI=0042
DS=1A04 ES=19F4 SS=1A04 CS=1A0C IP=0059 NV UP DI PL NZ NA PO NC
1A0C:0059 83FB01   CMP   BX,+01
-d1a04:ffe0
1A04:FFE0 B9 0B 0D 02 03 00 0D 02-61 00 59 00 0C 1A D4 16   9......a.Y...T.
1A04:FFF0 67 00 43 00 42 00 41 00-03 00 58 00 00 00 F4 19   g.C.B.A...X...t.
-t

AX=020D BX=0002 CX=0041 DX=000D SP=FFF0 BP=0061 SI=0043 DI=0042
DS=1A04 ES=19F4 SS=1A04 CS=1A0C IP=005C NV UP DI PL NZ NA PO NC
1A0C:005C 7418   JZ   0076
-t

AX=020D BX=0002 CX=0041 DX=000D SP=FFF0 BP=0061 SI=0043 DI=0042
DS=1A04 ES=19F4 SS=1A04 CS=1A0C IP=005E NV UP DI PL NZ NA PO NC
1A0C:005E E82C00   CALL   008D
-p

AX=020D BX=0002 CX=0041 DX=000D SP=FFE8 BP=0061 SI=0043 DI=0042
DS=1A04 ES=19F4 SS=1A04 CS=1A0C IP=0061 NV UP DI PL NZ NA PO NC
1A0C:0061 4B   DEC   BX
-d1a04:ffe0
1A04:FFE0 61 00 61 00 0C 1A D4 16-42 00 43 00 41 00 02 00   a.a...T.B.C.A...
1A04:FFF0 67 00 43 00 42 00 41 00-03 00 58 00 00 00 F4 19   g.C.B.A...X...t.
-t

AX=020D BX=0001 CX=0041 DX=000D SP=FFE8 BP=0061 SI=0043 DI=0042
DS=1A04 ES=19F4 SS=1A04 CS=1A0C IP=0062 NV UP DI PL NZ NA PO NC
1A0C:0062 87F7   XCHG   DI,SI
-t

AX=020D BX=0001 CX=0041 DX=000D SP=FFE8 BP=0061 SI=0042 DI=0043
DS=1A04 ES=19F4 SS=1A04 CS=1A0C IP=0064 NV UP DI PL NZ NA PO NC
1A0C:0064 E8F2FF   CALL   0059
-t

AX=020D BX=0001 CX=0041 DX=000D SP=FFE6 BP=0061 SI=0042 DI=0043
DS=1A04 ES=19F4 SS=1A04 CS=1A0C IP=0059 NV UP DI PL NZ NA PO NC
1A0C:0059 83FB01   CMP   BX,+01
-d1a04:ffe0
1A04:FFE0 59 00 0C 1A D4 16 67 00-42 00 43 00 41 00 02 00   Y...T.g.B.C.A...
1A04:FFF0 67 00 43 00 42 00 41 00-03 00 58 00 00 00 F4 19   g.C.B.A...X...t.
-t

AX=020D BX=0001 CX=0041 DX=000D SP=FFE6 BP=0061 SI=0042 DI=0043
DS=1A04 ES=19F4 SS=1A04 CS=1A0C IP=005C NV UP DI PL ZR NA PE NC

1A0C:005C 7418   JZ   0076
-t

AX=020D BX=0001 CX=0041 DX=000D SP=FFE6 BP=0061 SI=0042 DI=0043
DS=1A04 ES=19F4 SS=1A04 CS=1A0C IP=0076   NV UP DI PL ZR NA PE NC
1A0C:0076 E80100   CALL   007A
-p
A1C

AX=020D BX=0001 CX=0041 DX=000D SP=FFE6 BP=0061 SI=0042 DI=0043
DS=1A04 ES=19F4 SS=1A04 CS=1A0C IP=0079   NV UP DI PL NZ NA PE NC
1A0C:0079 C3   RET
-t

AX=020D BX=0001 CX=0041 DX=000D SP=FFE8 BP=0061 SI=0042 DI=0043
DS=1A04 ES=19F4 SS=1A04 CS=1A0C IP=0067   NV UP DI PL NZ NA PE NC
1A0C:0067 E82A00   CALL   0094
-d1a04:ffe0
1A04:FFE0 61 00 67 00 0C 1A D4 16-42 00 43 00 41 00 02 00   a.g...T.B.C.A...
1A04:FFF0 67 00 43 00 42 00 41 00-03 00 58 00 00 00 F4 19   g.C.B.A...X...t.
-p

AX=020D BX=0002 CX=0041 DX=000D SP=FFF0 BP=006A SI=0043 DI=0042
DS=1A04 ES=19F4 SS=1A04 CS=1A0C IP=006A   NV UP DI PL NZ NA PE NC
1A0C:006A E80D00   CALL   007A
-d1a04:ffe0
-d1a04:ffe0
1A04:FFE0 61 00 67 00 0C 1A 6A 00-6A 00 6A 00 0C 1A D4 16   a.g...j.j.j...T.
1A04:FFF0 67 00 43 00 42 00 41 00-03 00 58 00 00 00 F4 19   g.C.B.A...X...t.
-p
A2B

AX=020D BX=0002 CX=0041 DX=000D SP=FFF0 BP=006A SI=0043 DI=0042
DS=1A04 ES=19F4 SS=1A04 CS=1A0C IP=006D   NV UP DI PL NZ NA PE NC
1A0C:006D 4B   DEC   BX
-t

AX=020D BX=0001 CX=0041 DX=000D SP=FFF0 BP=006A SI=0043 DI=0042
DS=1A04 ES=19F4 SS=1A04 CS=1A0C IP=006E   NV UP DI PL NZ NA PO NC
1A0C:006E 87CE   XCHG   SI,CX
-t

AX=020D BX=0001 CX=0043 DX=000D SP=FFF0 BP=006A SI=0041 DI=0042
DS=1A04 ES=19F4 SS=1A04 CS=1A0C IP=0070   NV UP DI PL NZ NA PO NC
1A0C:0070 E8E6FF   CALL   0059
-t

AX=020D BX=0001 CX=0043 DX=000D SP=FFEE BP=006A SI=0041 DI=0042
DS=1A04 ES=19F4 SS=1A04 CS=1A0C IP=0059   NV UP DI PL NZ NA PO NC
1A0C:0059 83FB01   CMP   BX,+01
-d1a04:ffe0
1A04:FFE0 6A 00 04 1A 0D 02 6A 00-59 00 0C 1A D4 16 73 00   j.....j.Y...T.s.
1A04:FFF0 67 00 43 00 42 00 41 00-03 00 58 00 00 00 F4 19   g.C.B.A...X...t.
-t

AX=020D BX=0001 CX=0043 DX=000D SP=FFEE BP=006A SI=0041 DI=0042
DS=1A04 ES=19F4 SS=1A04 CS=1A0C IP=005C   NV UP DI PL ZR NA PE NC

```
1A0C:005C 7418              JZ      0076
-t

AX=020D  BX=0001  CX=0043  DX=000D  SP=FFEE  BP=006A  SI=0041  DI=0042
DS=1A04  ES=19F4  SS=1A04  CS=1A0C  IP=0076  NV UP DI PL ZR NA PE NC
1A0C:0076 E80100            CALL    007A
-p
C1B
AX=020D  BX=0001  CX=0043  DX=000D  SP=FFEE  BP=006A  SI=0041  DI=0042
DS=1A04  ES=19F4  SS=1A04  CS=1A0C  IP=0079  NV UP DI PL NZ NA PE NC
1A0C:0079 C3                RET
-t
AX=020D  BX=0001  CX=0043  DX=000D  SP=FFF0  BP=006A  SI=0041  DI=0042
DS=1A04  ES=19F4  SS=1A04  CS=1A0C  IP=0073  NV UP DI PL NZ NA PE NC
1A0C:0073 EB04              JMP     0079
-d1a04:ffe0
1A04:FFE0  04 1A B9 0B 02 01 0D 02-6A 00 73 00 0C 1A D4 16   ..9.....j.s...T.
1A04:FFF0  67 00 43 00 42 00 41 00-03 00 58 00 00 00 F4 19   g.C.B.A...X...t.
-t
AX=020D  BX=0001  CX=0043  DX=000D  SP=FFF0  BP=006A  SI=0041  DI=0042
DS=1A04  ES=19F4  SS=1A04  CS=1A0C  IP=0079  NV UP DI PL NZ NA PE NC
1A0C:0079 C3                RET
-t
AX=020D  BX=0001  CX=0043  DX=000D  SP=FFF2  BP=006A  SI=0041  DI=0042
DS=1A04  ES=19F4  SS=1A04  CS=1A0C  IP=0067  NV UP DI PL NZ NA PE NC
1A0C:0067 E82A00            CALL    0094
-d1a04:ffe0
1A04:FFE0  04 1A B9 0B 02 01 0D 02-0D 02 6A 00 67 00 0C 1A   ..9.......j.g...
1A04:FFF0  D4 16 43 00 42 00 41 00-03 00 58 00 00 00 F4 19   T.C.B.A...X...t.
-p
AX=020D  BX=0003  CX=0041  DX=000D  SP=FFFA  BP=006A  SI=0042  DI=0043
DS=1A04  ES=19F4  SS=1A04  CS=1A0C  IP=006A  NV UP DI PL NZ NA PE NC
1A0C:006A E80D00            CALL    007A
-d1a04:ffe0
1A04:FFE0  04 1A B9 0B 02 01 0D 02-0D 02 6A 00 67 00 0C 1A   ..9.......j.g...
1A04:FFF0  6A 00 6A 00 6A 00 0C 1A-D4 16 58 00 00 00 F4 19   j.j.j...T.X...t.
-p
A3C
AX=020D  BX=0003  CX=0041  DX=000D  SP=FFFA  BP=006A  SI=0042  DI=0043
DS=1A04  ES=19F4  SS=1A04  CS=1A0C  IP=006D  NV UP DI PL NZ NA PE NC
1A0C:006D 4B                DEC     BX
-t
AX=020D  BX=0002  CX=0041  DX=000D  SP=FFFA  BP=006A  SI=0042  DI=0043
DS=1A04  ES=19F4  SS=1A04  CS=1A0C  IP=006E  NV UP DI PL NZ NA PO NC
1A0C:006E 87CE              XCHG    SI,CX
-t
AX=020D  BX=0002  CX=0042  DX=000D  SP=FFFA  BP=006A  SI=0041  DI=0043
DS=1A04  ES=19F4  SS=1A04  CS=1A0C  IP=0070  NV UP DI PL NZ NA PO NC
1A0C:0070 E8E6FF            CALL    0059
```

-t
```
AX=020D  BX=0002  CX=0042  DX=000D  SP=FFF8  BP=006A  SI=0041  DI=0043
DS=1A04  ES=19F4  SS=1A04  CS=1A0C  IP=0059  NV UP DI PL NZ NA PO NC
1A0C:0059 83FB01          CMP       BX,+01
```
—d1a04:ffe0
```
1A04:FFE0  03 00 41 00 0D 00 42  00-43 00 6A 00 04 1A 0D 02   ..A...B.C.j.....
1A04:FFF0  6A 00 59 00 0C 1A D4  16-73 00 58 00 00 00 F4 19   j.Y...T.s.X...t.
```
—t
```
AX=020D  BX=0002  CX=0042  DX=000D  SP=FFF8  BP=006A  SI=0041  DI=0043
DS=1A04  ES=19F4  SS=1A04  CS=1A0C  IP=005C  NV UP DI PL NZ NA PO NC
1A0C:005C 7418            JZ        0076
```
—t
```
AX=020D  BX=0002  CX=0042  DX=000D  SP=FFF8  BP=006A  SI=0041  DI=0043
DS=1A04  ES=19F4  SS=1A04  CS=1A0C  IP=005E  NV UP DI PL NZ NA PO NC
1A0C:005E E82C00          CALL      008D
```
—p
```
AX=020D  BX=0002  CX=0042  DX=000D  SP=FFF0  BP=0061  SI=0041  DI=0043
DS=1A04  ES=19F4  SS=1A04  CS=1A0C  IP=0061  NV UP DI PL NZ NA PO NC
1A0C:0061 4B              DEC       BX
```
—d1a04:ffe0
```
1A04:FFE0  03 00 41 00 0D 00 42  00-61 00 61 00 0C 1A D4 16   ..A...B.a.a...T.
1A04:FFF0  43 00 41 00 42 00 02  00-73 00 58 00 00 00 F4 19   C.A.B...s.X...t.
```
—t
```
AX=020D  BX=0001  CX=0042  DX=000D  SP=FFF0  BP=0061  SI=0041  DI=0043
DS=1A04  ES=19F4  SS=1A04  CS=1A0C  IP=0062  NV UP DI PL NZ NA PO NC
1A0C:0062 87F7            XCHG      DI,SI
```
—t
```
AX=020D  BX=0001  CX=0042  DX=000D  SP=FFF0  BP=0061  SI=0043  DI=0041
DS=1A04  ES=19F4  SS=1A04  CS=1A0C  IP=0064  NV UP DI PL NZ NA PO NC
1A0C:0064 E8F2FF          CALL      0059
```
—t
```
AX=020D  BX=0001  CX=0042  DX=000D  SP=FFEE  BP=0061  SI=0043  DI=0041
DS=1A04  ES=19F4  SS=1A04  CS=1A0C  IP=0059  NV UP DI PL NZ NA PO NC
1A0C:0059 83FB01          CMP       BX,+01
```
—d1a04:ffe0
```
1A04:FFE0  03 00 41 00 0D 02 61  00-59 00 0C 1A D4 16 67 00   ..A...a.Y...T.g.
1A04:FFF0  43 00 41 00 42 00 02  00-73 00 58 00 00 00 F4 19   C.A.B...s.X...t.
```
—t
```
AX=020D  BX=0001  CX=0042  DX=000D  SP=FFEE  BP=0061  SI=0043  DI=0041
DS=1A04  ES=19F4  SS=1A04  CS=1A0C  IP=005C  NV UP DI PL ZR NA PE NC
1A0C:005C 7418            JZ        0076
```
—t
```
AX=020D  BX=0001  CX=0042  DX=000D  SP=FFEE  BP=0061  SI=0043  DI=0041
DS=1A04  ES=19F4  SS=1A04  CS=1A0C  IP=0076  NV UP DI PL ZR NA PE NC
1A0C:0076 E80100          CALL      007A
```
—p
B1A
```
AX=020D  BX=0001  CX=0042  DX=000D  SP=FFEE  BP=0061  SI=0043  DI=0041
```

```
DS=1A04   ES=19F4   SS=1A04   CS=1A0C   IP=0079    NV UP DI PL NZ NA PE NC
1A0C:0079 C3                  RET
-t

AX=020D   BX=0001   CX=0042   DX=000D   SP=FFF0   BP=0061   SI=0043   DI=0041
DS=1A04   ES=19F4   SS=1A04   CS=1A0C   IP=0067    NV UP DI PL NZ NA PE NC
1A0C:0067 E82A00              CALL      0094
-d1a04:ffe0
1A04:FFE0  04 1A B9 0B 02 01 0D 02-61 00 67 00 0C 1A D4 16    ..9.....a.g...T.
1A04:FFF0  43 00 41 00 42 00 02 00-73 00 58 00 00 00 F4 19    C.A.B...s.X...t.
-p

AX=020D   BX=0002   CX=0042   DX=000D   SP=FFF8   BP=006A   SI=0041   DI=0043
DS=1A04   ES=19F4   SS=1A04   CS=1A0C   IP=006A    NV UP DI PL NZ NA PE NC
1A0C:006A E80D00              CALL      007A
-d1a04:ffe0
1A04:FFE0  04 1A B9 0B 02 01 0D 02-61 00 67 00 0C 1A 6A 00    ..9.....a.g...j.
1A04:FFF0  6A 00 6A 00 0C 1A D4 16-73 00 58 00 00 00 F4 19    j.j...T.s.X...t.
-p
B2C

AX=020D   BX=0002   CX=0042   DX=000D   SP=FFF8   BP=006A   SI=0041   DI=0043
DS=1A04   ES=19F4   SS=1A04   CS=1A0C   IP=006D    NV UP DI PL NZ NA PE NC
1A0C:006D 4B                  DEC       BX
-t

AX=020D   BX=0001   CX=0042   DX=000D   SP=FFF8   BP=006A   SI=0041   DI=0043
DS=1A04   ES=19F4   SS=1A04   CS=1A0C   IP=006E    NV UP DI PL NZ NA PO NC
1A0C:006E 87CE                XCHG      SI,CX
-t

AX=020D   BX=0001   CX=0041   DX=000D   SP=FFF8   BP=006A   SI=0042   DI=0043
DS=1A04   ES=19F4   SS=1A04   CS=1A0C   IP=0070    NV UP DI PL NZ NA PO NC
1A0C:0070 E8E6FF              CALL      0059
-t

AX=020D   BX=0001   CX=0041   DX=000D   SP=FFF6   BP=006A   SI=0042   DI=0043
DS=1A04   ES=19F4   SS=1A04   CS=1A0C   IP=0059    NV UP DI PL NZ NA PO NC
1A0C:0059 83FB01              CMP       BX,+01
-d1a04:ffe0
1A04:FFE0  42 00 0D 00 41 00 43 00-6A 00 04 1A 0D 02 6A 00    B...A.C.j.....j.
1A04:FFF0  59 00 0C 1A D4 16 73 00-73 00 58 00 00 00 F4 19    Y...T.s.s.X...t.
-t

AX=020D   BX=0001   CX=0041   DX=000D   SP=FFF6   BP=006A   SI=0042   DI=0043
DS=1A04   ES=19F4   SS=1A04   CS=1A0C   IP=005C    NV UP DI PL ZR NA PE NC
1A0C:005C 7418                JZ        0076
-t

AX=020D   BX=0001   CX=0041   DX=000D   SP=FFF6   BP=006A   SI=0042   DI=0043
DS=1A04   ES=19F4   SS=1A04   CS=1A0C   IP=0076    NV UP DI PL ZR NA PE NC
1A0C:0076 E80100              CALL      007A
-p
A1C

AX=020D   BX=0001   CX=0041   DX=000D   SP=FFF6   BP=006A   SI=0042   DI=0043
DS=1A04   ES=19F4   SS=1A04   CS=1A0C   IP=0079    NV UP DI PL NZ NA PE NC
```

1A0C:0079 C3   RET
-t

AX=020D BX=0001 CX=0041 DX=000D SP=FFF8 BP=006A SI=0042 DI=0043
DS=1A04 ES=19F4 SS=1A04 CS=1A0C IP=0073 NV UP DI PL NZ NA PE NC
1A0C:0073 EB04   JMP   0079

-d1a04:ffe0
1A04:FFE0 0D 00 42 00 43 00 6A 00-04 1A B9 0B 02 01 0D 02   ..B.C.j...9.....
1A04:FFF0 6A 00 73 00 0C 1A D4 16-73 00 58 00 00 00 F4 19   j.s...T.s.X...t.
-t

AX=020D BX=0001 CX=0041 DX=000D SP=FFF8 BP=006A SI=0042 DI=0043
DS=1A04 ES=19F4 SS=1A04 CS=1A0C IP=0079 NV UP DI PL NZ NA PE NC
1A0C:0079 C3   RET
-t

AX=020D BX=0001 CX=0041 DX=000D SP=FFFA BP=006A SI=0042 DI=0043
DS=1A04 ES=19F4 SS=1A04 CS=1A0C IP=0073 NV UP DI PL NZ NA PE NC
1A0C:0073 EB04   JMP   0079

-d1a04:ffe0
1A04:FFE0 0D 00 42 00 43 00 6A 00-04 1A B9 0B 02 01 0D 02   ..B.C.j...9.....
1A04:FFF0 0D 02 6A 00 73 00 0C 1A-D4 16 58 00 00 00 F4 19   ..j.s...T.X...t.
-t

AX=020D BX=0001 CX=0041 DX=000D SP=FFFA BP=006A SI=0042 DI=0043
DS=1A04 ES=19F4 SS=1A04 CS=1A0C IP=0079 NV UP DI PL NZ NA PE NC
1A0C:0079 C3   RET
-t

AX=020D BX=0001 CX=0041 DX=000D SP=FFFC BP=006A SI=0042 DI=0043
DS=1A04 ES=19F4 SS=1A04 CS=1A0C IP=0058 NV UP DI PL NZ NA PE NC
1A0C:0058 CB   RETF

-d1a04:ffe0
1A04:FFE0 0D 00 42 00 43 00 6A 00-04 1A B9 0B 02 01 0D 02   ..B.C.j...9.....
1A04:FFF0 0D 02 0D 02 6A 00 58 00-0C 1A D4 16 00 00 F4 19   ....j.X...T...t.
-t

AX=020D BX=0001 CX=0041 DX=000D SP=0000 BP=006A SI=0042 DI=0043
DS=1A04 ES=19F4 SS=1A04 CS=19F4 IP=0000 NV UP DI PL NZ NA PO NC
19F4:0000 CD20   INT   20
-p

Program terminated normally
-q

图 2.32 例 2.8 在 N=3 时的运行过程及堆栈状态

```
            ;—————————————————————————————
            ;—————————————
0059        hanoi   proc    near            ;define subprocedure
            ; Solves tower of HANOI puzzle.
            ; Argement: (BX)=N, (CX)=X, (SI)=Y, (DI)=Z.
0059  83 FB 01      cmp     bx,1            ;if N=1,execute basis
005C  74 18         je      basis
005E  E8 008D R     call    save            ;save N,X,Y,Z
0061  4B            dec     bx
```

0062	87 F7		xchg	si,di	
0064	E8 0059 R		call	hanoi	;call HANOI (N-1,X,Z,Y)
0067	E8 0094 R		call	restor	;restore N,X,Y,Z
006A	E8 007A R		call	print	;print XNZ
006D	4B		dec	bx	
006E	87 CE		xchg	cx,si	
0070	E8 0059 R		call	hanoi	;call HANOI (N-1,Y,X,Z)
0073	EB 04 90		jmp	return	
0076	E8 007A R	basis:	call	print	;print X1Z
0079	C3	return:	ret		;return
007A			hanoi	endp	;end subprocedure

;————————————————————————————————

图 2.33　例 2.8 中 hanoi 子程序的 LST 清单

② debug 所显示的汇编指令是反汇编得到的结果,因而只显示十六进制地址而无符号名,为便于读者阅读,给出对应关系如下:

　　CALL　　008D　　为　　CALL　　SAVE
　　CALL　　0059　　为　　CALL　　HANOI
　　CALL　　007A　　为　　CALL　　PRINT
　　CALL　　0094　　为　　CALL　　RESTOR

③ 在用 d 命令显示的堆栈内容中,位于堆栈底部的内容为:

　　1A04:FFFA　　　　0058
　　1A04:FFFC　　　　0000
　　1A04:FFFE　　　　19F4

其中 0058 为主程序 main 调用 hanoi 子程序时保存的返回地址,在程序运行的最后将通过它返回主程序去执行地址为 0058 的 RET 指令。而 19F4:0000 为机器的操作系统 DOS 调用本程序时的返回地址,在机器执行 0058 的 RET 指令时,将通过它返回 DOS 去执行 19F4:0000 单元中的指令 INT　20 以结束本程序的执行。

二、实验题

实验 2.4　查找电话号码

1. 题目:查找电话号码 phone

2. 实验要求:

(1) 要求程序建立一个可存放 50 项的电话号码表,每项包括人名(20 个字符)及电话号码(8 个字符)两部分;

(2) 程序可接收输入人名及相应的电话号码,并把它们加入电话号码表中;

(3) 凡有新的输入后,程序应按人名对电话号码表重新排序;

(4) 程序可接收需要查找电话号码的人名,并从电话号码表中查出其电话号码,再在屏幕上以如下格式显示出来。

　　　　name　　　　　　tel.
　　　　××××　　　　××××

3. 提示:程序采用子程序结构。主程序的主要部分如下:

- 显示提示符'Input name:';
- 调用子程序 input_name 接收人名;
- 调用子程序 stor_name 把人名存入电话号码表 tel_tab 中;
- 显示提示符'Input a telephone number:';
- 调用子程序 inphone 接收电话号码,并把它存入电话号码表 tel_tab 中;
- 如输入已结束则调用 name_sort 子程序对电话号码表按人名排序;
- 显示提示符'Do you want a telephone number? (Y/N)';
- 回答 N 则退出程序;
- 回答 Y 则再显示提示符'name? ';
- 调用子程序 input_name 接收人名;
- 调用子程序 name_search 在电话号码表中查找所要的电话号码;
- 调用子程序 printline 按要求格式显示人名及电话号码;
- 重复查号提示符直至用户不再要求查号为止。

4. 实验报告要求:

除 1.1 节中提出的要求外,还应增加以下内容:
- 画出模块层次图;
- 写出各模块说明。

实验 2.5[*]　求 Fibonacci 数

1. 题目:求 Fibonacci 数 fib
2. 实验要求:

(1) 程序接收由用户键入的范围在 0～100(不包括 0 和 100)之间的 n 值;

(2) 根据给定的 n 值,计算 Fibonacci 数。

其定义如下:

$$\begin{cases} FIB(1)=1 \\ FIB(2)=1 \\ FIB(n)=FIB(n-2)+FIB(n-1) \quad (n>2) \end{cases}$$

(3) 程序输出 FIB(n)值

3. 提示:

程序可由三部分组成:

(1) 输入部分:程序接收用户键入的 n 值,经十化二后存入 num 单元中;

(2) 用递归子程序 fibp 求出 FIB(n)值,结果存放在 result 单元中。在递归子程序的设计中,每次调用可把一帧信息存入堆栈,帧结构如下:

```
frame       struc
 save_bp    dw    ?
 save_ip    dw    ?
 num        dw    ?
 temp_addr  dw    ?
 result_addr dw   ?
```

```
        frame     ends
```
其中 temp 单元用来保存中间结果；

(3) 输出部分：把 result 中的运行结果经二化十后在屏幕上显示出来。

4. 实验报告要求：

除 1.1 节中所提出的要求外，还应增加以下内容：

- 画出模块层次图；
- 写出各模块说明；
- 画出当 n=5 时，堆栈最满时的堆栈情况图。

第三章 I/O 程序设计

3.1 发声系统程序设计

为了具有音响输出的能力,系统板上装有一个 $2\frac{1}{4}$ 英寸的扬声器以及控制电路和驱动电路。控制电路能以位触发和定时器控制两种不同的方式驱动扬声器发声。

(1) 位触发方式

程序直接控制 PPI(8255A 可编程序外围接口芯片)的输出控制寄存器(I/O 端口为 61H)的第 1 位,使该位按所需的频率进行 1 和 0 的交替变化,从而控制开关电路产生一串脉冲波形,这些脉冲经放大后驱动扬声器发出声音。如果控制这一串脉冲波形的脉宽和长度就可以产生不同频率和不同音长的声音。

采用位触发方式发声的程序段如下:

```
         in    al,   61H
         mov   ah,   al
         and   al,   11111100b    ;关断定时器通道 2 的门控
sound:   xor   al,   2            ;触发 61H 端口第 1 位
         out   61H,  al
         mov   cx,   dx           ;(dx)=控制脉宽的计数值
wait:    loop  wait                ;延时循环
         dec   bx                 ;(bx)=脉冲持续的时间
         jnz   sound
         mov   al,   ah
         out   61h,  al           ;恢复 61H 端口
```

(2) 利用定时器产生声音

这是利用机器硬件即 INTEL 8253/8254 定时器产生声音的一种方法。

CPU 通过对定时器的通道 2(端口地址为 42H)进行编程,使其 I/O 寄存器接收一个控制声音频率的 16 位计数值,端口 61H 的最低位控制通道 2 门控的开断,以产生特殊的音响。当定时器接收的计数值为 533H 时,能产生 896Hz 的声音,因此产生其它频率(Freq)的计数值就可由下式计算出来:

$$533H \times 896 \div Freq = 1234DCH \div Freq$$

在送出频率计数值之前,还要给方式寄存器(其端口地址为 43H)送一个方式值,也称为幻数,这个幻数决定对哪一个通道编程,采用什么模式,送入通道的计数值是一字节还是两字节,是二进制码还是 BCD 码。其位组合的格式如下:

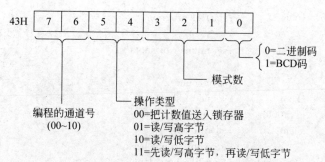

当通道2用于发声时,一般采用模式3,在模式3下,输出线为"1"和为"0"的时间各占计数时间的一半,因而产生一系列间隔均匀的脉冲。

下面是利用定时器产生指定频率声音的程序段:

```
        mov     al,     10110110b       ;位组合格式:通道2,两字节计数,模式3,二进制码
        out     43h,    al
        mov     dx,     12h
        mov     ax,     34Dch
        div     di                      ;(di)=freq
        out     42h,    al              ;频率计数值送通道2
        mov     al,     ah
        out     42h,    al
```

控制音长的时间可以简单地通过反复执行循环指令来得到。我们已知执行2801次LOOP指令约需10ms的时间,因此用10ms的倍数值来控制扬声器开关的时间间隔,就可控制音长。具体实现的指令序列如下:

```
        in      al,     61h
        mov     ah,     al
        or      al,     3
        out     61h,    al              ;接通扬声器
leng:   mov     cx,     2801            ;10ms音长的计数值
delay:  loop    delay
        dec     bx                      ;(bx)=10ms的倍数值
        jnz     leng
        mov     al,     ah
        out     61h,    al              ;关闭扬声器
```

有关计算机发声的原理,请参阅清华大学出版社1991年出版的《IBM-PC 汇编语言程序设计》第十一章"发声系统的程序设计"。

一、示例

例3.1 枪声程序 gun

```
TITLE   GUN---Makes machine gun sound
;               fires fixed number of shots
;****************************************
prognam segment                 ;define code segment
;----------------------------------------
main    proc    far             ;main part of program
        assume  cs : prognam
```

```
            org     100h            ;start of program
start:
            mov     cx,50d          ;set number of shots
new_shot:
            push    cx              ;save count
            call    shoot           ;sound of shot
            mov     cx,4000h        ;set up silent delay
silent:     loop    silent          ;silent delay
            pop     cx              ;get shots count back
            loop    new_shot        ;loop till shots done
            mov     al,48h
            out     61h,al          ;reset output port
            int     20h             ;return to DOS
main        endp                    ;end of main part of program
;————————————————————————————————————
;SUBROUTINE TO MAKE BRIEF NOISE
shoot       proc    near
            mov     dx,140h         ;initial value of wait
            mov     bx,20h          ;set count
            in      al,61h          ;get port 61
            and     al,11111100b    ;AND off bits 0,1
sound:      xor     al,2            ;toggle bit #1 in AL
            out     61h,al          ;output to port 61
            add     dx,9248h        ;add random pattern
            mov     cl,3            ;set to rotate 3 bits
            ror     dx,cl           ;rotate it
            mov     cx,dx           ;put in CX
            and     cx,1ffh         ;mask off upper 7 bits
            or      cx,10           ;ensure not too short
wait:       loop    wait            ;wait
;made noise long enough?
            dec     bx              ;done enough?
            jnz     sound           ;jump if not yet
;turn off sound
            and     al,11111100b    ;AND off bits 0,1
            out     61h,al          ;turn off bits 0,1
            ret                     ;return from subr
shoot       endp
;————————————————————————————————————
            prognam ends            ;end of code segment
;********************************************
            end     start           ;end assembly
```

图 3.1 例 3.1 的程序清单

运行这个程序,计算机就连续发出一串"枪声"。程序中驱动发声的部分采用了方法(1)位触发的原理,即通过不断变换端口 61H 第 1 位的值来控制开关电路发出一串脉冲,而控制脉冲宽度的计数值(在 cx 中)是一个 10~500 的随机数,因此,发出的"枪声"频率不同,听起来时缓时急,非常逼真。

在发声操作结束之后,返回 DOS 之前有两条指令完成输出端口的复位操作:

```
            mov     al,         48h
            out     61h,        al
```

例 3.2 演奏音阶程序 musex

```
TITLE MUSEX -- tone of do_ re_ mi
;————————————————————————————
dseg    segment para 'data'
        dw 0
mus_f   dw 262,294,330,349,392,440,494,523,0
mus_t   dw 7 dup(50),100
flag    dw -1
dseg ends
;————————————————————————————
cseg    segment para 'code'
assume  cs:cseg, ds:dseg
tone    proc    far
        push    ds              ;put return addr.
        sub     ax,ax           ;  on stack
        push    ax
        mov     ax,dseg         ;initialize DS
        mov     ds,ax

        lea     si,mus_f        ;freq table offset in SI
next:   lea     bp,mus_t        ;time table offset in BP
freq:   mov     di,[si]         ;read next frequency
        cmp     di,0            ;end of tone ?
        je      end_f           ;yes,one times is end
        mov     bx,ds:[bp]      ;no,fetch the duration

        mov     al,0b6h         ;put magic number
        out     43h,al          ;    into timer2
        mov     dx,12h
        mov     ax,533h*896
        div     di              ;(DX,AX)/(DI)
        out     42h,al          ;LSB into timer2
        mov     al,ah
        out     42h,al          ;MSB into timer2
        in      al,61h
        mov     ah,al
        or      al,3            ;turn speaker on
        out     61h,al
long:   mov     cx,2801         ;length of tone
delay:  loop    delay
        dec     bx
        jnz     long
        mov     al,ah           ;recover the port
        out     61h,al

        test    flag,1          ;raise or lower ?
        jz      lower
        add     si,2            ;inc the freq pointer
        jmp     cont
```

```
lower:  sub     si,2            ;dec the freq pointer
cont:   add     bp,2            ;update mus_t pointer
        jmp     freq            ;go process next note
end_f:  inc     flag
        jnz     exit            ;two times end
        sub     si,2
        jmp     next            ;second times
exit:   mov     al,48h          ;reset port
        out     61h,al
        ret                     ;return to DOS
tone    endp
cseg    ends
;————————————————————————
        end     tone
```

图 3.2 例 3.2 的程序清单

运行这个程序,计算机从 1 到 1,又从 1 到 1 反复两次奏出一个音阶。发声的方法是利用定时器控制声音脉冲的方法,它包括三步:首先,在定时器 2 的方式寄存器中(I/O 端口为 43H)装入一个幻数 0B6H,以进行初始化。其次给定时器 2 的 42H 端口送入一个控制声音频率的数据((533H×896)/音符的频率值)。最后打开扬声器的与门,使之发出一定频率的声音。

为此,在程序中的数据段设置了两个数据区,mus_f 数据区存放着一个音阶的频率值,在 mus_t 数据区存放着 8 个音符发声的持续时间值(10ms 的倍数),这两组数据分别控制定时器 2 及扬声器开关的时间,以产生不同音调和音长的声音。

数据段中还有一个变量 flag,初始值为 -1,当一个音阶的 8 个音符演奏完后(0 值为结束标志),flag 加 1,接着再从 mus_f 中反向取出 8 个音符的频率值送定时器 2,0 值仍为反向取频率值的结束标志。当 flag 加 1 不为 0 时,程序完成输出控制寄存器 61H 的复位操作后返回 DOS。程序中如没有复位操作的指令,则在程序退出运行返回 DOS 之后,系统将停止一切工作。

本例使用定时器控制发声,它只在一个八度的音阶范围内演奏,但只要扩大音阶范围,并把频率值和延迟时间按一定的规律排列,就可形成多种变化的音响效果。

二、实验题

实验 3.1*   乐曲程序(1)

1. 题目:乐曲程序(1)
2. 实验要求:采用位触发方式编写程序,使计算机发出音响并奏出《两只老虎》的乐曲。

乐曲《两只老虎》的简谱如下:

<pre>
                    两只老虎

            1=C   4/4

            1 2 3 1 | 1 2 3 1 | 3 4 5 — |
            3 4 5 — | 56 54 3 1 | 56 54 3 1 |
            2 5 1 — | 2 5 1 — |
</pre>

3. 提示

(1) 一首乐曲是由不同频率和节拍的音调组成的，因此控制驱动脉冲的频率和持续时间就是编写乐曲程序的关键。

下表为两个八度的音阶表：

音名	C	D	E	F	G	A	B
音符	1.	2.	3.	4.	5.	6.	7.
频率	131	147	165	175	196	220	247

音名	C′	D′	E′	F′	G′	A′	B′	C″
音符	1	2	3	4	5	6	7	1̇
频率	262	294	330	349	392	440	494	523

(2) 利用位触发方式演奏乐曲，程序必须将音符的频率转化为控制脉冲宽度的计数值，其原理如下图所示：

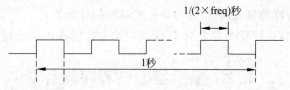

设音频为 freq，则脉冲周期为 $1/freq$，一个半波形（脉宽）所需的时间就为 $1/(2 \times freq)$ 秒，用这个时间值来控制输出端口 61H 的第 1 位 1/0 的延迟时间，也就是控制了开关电路所产生脉冲的频率，从而驱动扬声器产生一定频率的音响。

$1/(2 \times freq)$ 秒的延迟时间可简单地通过 loop 指令的循环来取得，我们知道 2801 次 loop 指令循环执行所需时间是 10ms，所以 1 秒钟时间约执行 2801×100 次 loop 指令。

$$1/(2 \times freq) = (2801 \times 100)/(2 \times freq) = (2801 \times 50)/freq$$

根据上式控制脉宽的计数值可由以下指令来实现：

```
mov     ax,2801
mov     bx,50
mul     bx
div     di          ;(di)=freq
mov     dx,ax       ;(dx)=1/(2*freq)
```

(3) 乐曲中的节拍决定了各音频持续的时间，如二分音符为 2 拍，持续时间取 0.5 秒，(50×10ms)，我们就把 10ms 的倍数 50 作为 2 拍的计数值，同理，1 拍的计数值为 25，1/2 拍的计数值为 12……。这样，按照一首乐曲的曲谱将各音符的频率和节拍计数值定义成两个数据表，作为程序控制发声的数据。对于位触发式的发声方法，还需将节拍计数值再扩大 6~10 倍：

```
mov     ax,8
mul     bx
mov     bx,ax       ;(bx)=扩大8倍的节拍计数值
```

(4) 乐曲演奏完后,程序在返回 DOS 之前,输出控制寄存器必须进行复位操作。对 PC 机,向 61H 端口送控制值 48h,对其它 PC 兼容机送出控制值 88h 或 44h。

实验 3.2 乐曲程序(2)

1. 题目:乐曲程序(2)

2. 实验要求:利用定时器产生声音的方法编写程序,使 PC 机奏出《两只老虎》的乐曲。

3. 提示:

(1) 乐谱中的每个音符具有音高和音长两种属性,因此和实验 3.1 一样,按照乐谱将每个音符的频率和节拍定义成两个数据表,程序从频率表中取出一个音符的频率值以产生一定音高的声音,同时取出相应的节拍计数值以控制这一频率的声音延长的时间。

(2) 每个音符的频率值 freq 经过转换后送入定时器的 42H 端口,以产生相应频率的脉冲。转换的公式为:
$$533H \times 896 \div freq = 1234DCH \div freq$$

(3) 节拍时间表中存放的计数值是音长的 1/10,如 2 拍的音长定为 0.5 秒,对应的计数值应为 50,1 拍的音长是 0.25 秒,对应的计数值应为 25。然后控制 loop 指令反复执行 2801×n 次来取得音符的延长时间(n 为音长所对应的计数值)。

(4) 乐曲演奏完后,同样要完成对 61H 端口的复位操作以避免死机。

3.2 显示器 I/O 程序设计

IBM 微型机的视频系统都是以 6845 CRT 控制器或以基于 6845 的定制芯片为核心构成的。6845 CRT 控制器的基本任务是:(1)将 CRT 工作方式设置成字符或图形方式之一;(2)对 ASCII 代码号进行译码,并从适配器 ROM(有时从 RAM)芯片中取出对应字符的数据;(3)对数据进行解码,以得出属性或彩色并依此调整屏幕显示;(4)生成并控制光标。

PC 的视频系统都有用于反映屏幕图像数据的显示缓冲区,通过扫描缓冲区中的数据,在屏幕上显示出相应的字符或图形,当显示缓冲区里存有多幅图象时,我们称每幅图象为一页。对不同的适配器,显示缓冲区的大小及在内存中的位置不同。对单色适配器,其显示缓冲区有 4K 字节的容量,内存空间地址从 B000:0000 开始。彩色图形适配器有 16K 字节的存储容量,内存起始地址为 B800:0000。PC_{jr} 以常规的 RAM 作为视频缓冲区,在 BIOS 初始化系统时,现有存储容量的前 16K 被定义为显示缓冲区。EGA 可配有 64K、128K 或 256K RAM,除了作为视频缓冲区外,该显存也可保存多达 1024 个字符的数据。其高级图形方式的起始地址为 A000:0000,标准单色和彩色图形方式的起始地址则分别为 B000:0000 和 B800:0000。

各类视频系统在显示文本时,工作方式是相同的,存储器总共分配了 4000 个字节,对应屏幕上的 2000 个字符位置(25 行×80 列),每个字符位置对应 2 个字节,其中低地址字节内是 ASCII 代码,高地址字节则保存着字符显示特性的信息(属性)。

彩色存储从 B800:0000 开始的 8K 存储器单元对应于屏幕上的偶数行,从 B800:

2000h 开始的另外 8K 字节对应着屏幕的奇数行。对 320×200 彩色图形显示方式,全屏幕共有 64000 个像素,每 4 个像素用一个字节来表示,表示一个像素的二位可表示四种不同的颜色。

在屏幕上显示字符或图形的方法很多,但归根结底,微机所做的工作就是在显示存储区的特定位置处放入字符数据(ASCII 码及属性)或像素值。通过编程将字符数据或像素值直接置入显示存储区,这种方法称为"存储器映射"。

在文本方式下,屏幕上的字符位置所对应的显示缓冲区的偏移地址为:

行号(0～24)×160＋列号(0～79)×2

如在屏幕的中心行(12 行)显示一行"＊"号:

```
        mov   ax,    0B000h      ;单色显示缓冲区
        mov   es,    ax
        mov   bh,    12          ;行号
        mov   bl,    0           ;列号
        mov   cx,    80          ;80 个字符
        mov   al,    160         ;
        mul   bh                 ;(ax)=行号×160
        rol   bl,    1           ;列号×2
        mov   bh,    0
        add   bx,    ax          ;(bx)=显存地址
        mov   al,    '*'         ;ASCII 码
        mov   ah,    07          ;属性
next:   mov   es:[bx],al         ;字符送入低字节
        mov   es:[bx+1],ah       ;属性送入高字节
        inc   bx                 ;
        inc   bx                 ;修改地址
        loop  next               ;在下一个位置显示字符
```

在图形方式下,屏幕上的一个象素所对应的存储器地址为:

显示缓冲区偏址(0 或 2000h)＋行号/2×80＋列号/4

偶数行的缓冲区偏址为 0,奇数行的缓冲区偏址为 2000h。对 320×200 显示方式,每行 320 个像素共需用 80 个存储器字节来表示,列号的低两位指出像素在字节中的位组号。按上式计算出来的偏移地址的字节内可存放四个像素值:

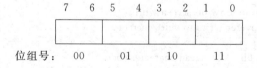

所以屏幕位置上的像素值还要按其位组号放入已知字节中相应的两位中。

下面的程序完成在屏幕 100 行,180 列处显示一个点(像素值为 2):

```
        mov   ax,    0b800h      ;彩色显示存储区
        mov   es,    ax
        mov   cx,    100         ;行号
        mov   dx,    180         ;列号
        test  cl,    1           ;是奇数行吗
        jz    even_row           ;是偶数行
```

```
            mov     bx,       2000h         ;奇数行偏址
            jmp     short     continue
    even_row: mov   bx,       0             ;偶数行偏址
    continue: shr   cx,       1             ;行号除以2
            mov     al,       80            ;每行80字节
            mul     cl                      ;(ax)=当前行之前的字节数
            push    dx                      ;保存列号
            shr     dx,       1             ;列号/4
            shr     dx,       1
            add     ax,       dx            ;(ax)=行号/2×80+列号/4
            add     bx,       ax            ;(bx)=显存偏址+(ax)
            pop     dx                      ;恢复列号
            mov     cx,       dx            ;
            not     cl                      ;位组号求反
            and     cl,       00000011b     ;位组号转换为
            shl     cl,       1             ;移位次数
            mov     ah,       es:[bx]       ;取出像素所在的字节
            ror     ah,       cl            ;像素对应的位组移到低位
            and     ah,       11111100b     ;清除低两位
            mov     al,       pixel         ;pixel为像素值
            or      ah,       al            ;置入像素值
            rol     ah,       cl            ;移回原位组
            mov     es:[bx], ah             ;写入显存
```

采用存储器映射方法与使用操作系统提供的 DOS 和 BIOS 显示功能调用相比,前者的编程量稍大些,但因为能极大地提高屏幕显示速度,同时由于 BIOS 及 DOS 中断处理彩色方式的功能较弱,所以存储器映射法也是程序员经常采用的一种编程方法。然而,为了保持各种 PC 间的兼容性,在无特殊要求的情况下,应尽量使用 DOS 和 BIOS 的显示功能编写程序。

有关显示器程序设计的内容,可参阅教材第九章和第十章。

一、示例

例 3.3　光标轨迹程序 draw

```
TITLE DRAW -- Program to draw on screen with
    ; cursor arrows,character write to video memory
    ;----------------------------------------------------
        read_c      equ     0h      ;read character code
        key_rom     equ     16h     ;ROM keyboard routine
        up          equ     48h     ;scan code for up arrow
        down        equ     50h     ;scan code for down arrow
        right       equ     4dh     ;scan code for right arrow
        left        equ     4bh     ;scan code for left arrow
        block       equ     0dbh    ;solid graphics character
        esc         equ     1bh     ;escape key
    ;*************************************************************
        video       segment at 0b800h   ;define extra seg
        wd_buff     label   word
```

```
            v_buff      db         25 * 80 * 2 dup (?)
            video       ends
;************************************************
            pro_nam segment                   ;define code segment
;————————————————————————————————————————————————
            main        proc       far        ;main part of program
            assume      cs:pro_nam, es:video
;set up stack for return
start:      push        ds                    ;save DS
            sub         ax,ax                 ;set AX to zero
            push        ax                    ;put it on stack
;set ES to extra segment
            mov         ax,video
            mov         es,ax
;clear screen by writing zeros to it
; even bytes get 0 (character)
; odd bytes get 7 (normal "attribute")
            mov         cx,80*25              ;count
            mov         bx,0                  ;start of buff
clear:      mov         es:[wd_buff+bx],0700h
            inc         bx                    ;incr pointer
            inc         bx                    ; twice
            loop        clear                 ;do again
;screen pointer will be in CX register
; row    number (0 to 24d) in CH
; column number (0 to 79d) in CL
;set screen pointer to center of screen
            mov         ch,12d                ;# rows divided by 2
            mov         cl,40d                ;# columns div by 2
;get character from keyboard,using ROM BIOS routine
get_char:
            mov         ah,read_c             ;code for read char
            int         key_rom               ;keyboard I/O ROM call
            cmp         al,esc                ;is it escape ?
            jz          exit                  ;yes
            mov         al,ah                 ;put scan code in AL
            cmp         al,up                 ;is it UP arrow?
            jnz         not_up                ;no
            dec         ch                    ;yes , decrement row
not_up:
            cmp         al,down               ;is it DOWN arrow?
            jnz         not_down              ;no
            inc         ch                    ;yes, increment row
not_down:
            cmp         al,right              ;is it RIGHT arrow?
            jnz         not_right             ;no
            inc         cl                    ;yes,increment column
not_right:
            cmp         al,left               ;is it LEFT arrow?
            jnz         lite_it               ;no
            dec         cl                    ;yes,decrement column
lite_it:
```

```
            mov     al,160d         ;bytes per row into AL
            mul     ch              ;time # of rows
                                    ; result in AX
            mov     bl,cl           ;# of columns in BL
            rol     bl,1            ;times 2 to get bytes
            mov     bh,0            ;clear top part of BX
            add     bx,ax           ;add AX to it
                                    ; gives address offset
;address offset in BX . Put block char there
            mov     al,block
            mov     es:[v_buff + bx],al
            jmp     get_char        ;go get next arrow
exit:       ret                     ;return to DOS
main        endp                    ;end of main part of program
;————————————————————————————————————————————
pro_nam     ends                    ;end of code segment
;********************************************
            end     start           ;end assembly
```

图 3.3 例 3.3 的程序清单

例 3.3 所描述的程序能在屏幕上画出光标移动的轨迹。光标的移动受键盘上 ↑ ↓ ← → 光标键的控制,如果键入 Esc 键,则程序结束,返回 DOS。

该程序首先定义了一个附加段 video,并用 at 伪操作将该段的段地址指定在 0B800H。0B800H 是彩色图形适配器的显示缓冲区,如果使用单色显示器,显示缓冲区的段地址应为 0B00H。在汇编语言程序中,常把一些专用的数据区定义为附加段,而把其他的变量或信息存放在数据段,这样做一方面可保证专用数据区的存储容量,另一方面在数据段中修改数据或增减显示信息更方便一些。前面已提到 25×80 显示方式需要 4000 个字节的存储区,这是以字(wd_buff)和字节(v_buff)两种单位来定义同一个存储区,这是因为程序中在清除屏幕时,向所有单元写入同一数据,这时以字方式操作要简单、快速得多,而往显示存储区中"写"字符时,需要以字节为单位进行操作。显示存储区的容量以表达式 25×80×2 来表示,其意义比直接用 4000 更为明确。它说明字符显示是以 25×80 的方式,且每个字符分别由 2 个字符来表示其 ASCII 码和属性。

整个程序可分为三个功能段。第一段完成清除屏幕的工作,程序从偏移地址 0 开始,向显存的所有字单元中写入数据 0700H,其中 07H 是正常属性代码,00 是 ASCII 码,这种直接向显存单元写零来清除屏幕的方法是最直接最快速的清屏方法。

第二段程序用 BIOS INT 16H 功能从键盘读取字符,然后根据按动的光标控制键增减行号(在 CH 中)或列号(在 CL 中)。

程序的第三部分,在当前行号和列号的位置写入方块字符(ASCII 码为 BDH)。在写入字符之前先要将屏幕上的行号和列号转换成该字符位置所对应的显存地址,这就使用了转换公式:

$$(bx)=行号\times 80\times 2+列号\times 2$$

然后利用间接寻址方式直接将方块符 1BH 写入该地址的存储单元中:

```
            mov     al,1bh
```

        mov   es:[v_buff+bx],al

这时在光标指示的屏幕位置上立即显示出一个正常属性的方块符。

运行这个程序时,我们任意按动↑↓←→光标键,光标可围绕屏幕移动,其移动的轨迹就由一条方块符组成的粗线保留下来了。如果按照一定的轨迹移动光标键,还能组成一些简单的图案。图3.4就是在光标键的控制下,程序在屏幕上画出的"城墙"图案。

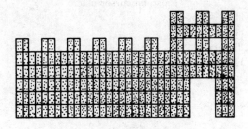

图 3.4　由方块符组成的"城墙"

例 3.4　窗口控制程序 wdex

下面的窗口控制程序所完成的工作和鼠标器控制屏幕窗口的功能相似。首先清除屏幕,紧接着在屏幕中心生成一个小窗口,其左上角的行列坐标为(10,30),右下角坐标为(15,40),光标定位在(0,0)。当按动光标控制键时,光标能上下左右移动,当光标移动到指定的位置上后,按下 END 键,此时光标的最后位置或作为窗口左上角坐标,或作为右下角坐标扩大或缩小原来的屏幕窗口。

图 3.5 是窗口控制程序的清单。

```
TITLE WDEX --- Variable_size video window
;──────────────────────────────────
dseg      segment
cury      db       0         ;current row #
curx      db       0         ;current column #
luy       db       10        ;upper_left row #
lux       db       30        ;upper_left col #
rdy       db       15        ;lower_right row #
rdx       db       40        ;lower_right col #
cont      db       5         ;row num of window
dseg      ends
;──────────────────────────────────
clear     macro              ;clear screen macro
          mov      ah,6
          mov      al,0
          mov      ch,0
          mov      cl,0
          mov      dh,24
          mov      dl,79
          mov      bh,7
          int      10h
          endm
scroll    macro    cont      ;make a window macro
          mov      ah,6
          mov      al,cont
```

```
                mov     ch,luy
                mov     cl,lux
                mov     dh,rdy
                mov     dl,rdx
                mov     bh,70h
                int     10h
                endm
postc           macro                   ;position cursor macro
                mov     ah,2
                mov     dh,cury
                mov     dl,curx
                mov     bh,0
                int     10h
                endm
;------------------------------------------------
cseg            segment
assume          cs:cseg, ds:dseg
main            proc    far
                push    ds              ;save for return
                sub     ax,ax
                push    ax
                mov     ax,dseg         ;set data segment addr.
                mov     ds,ax

                clear                   ;clear screen
                scroll  cont            ;make a window in centrn
                postc                   ;position cursor at(0,0)
input:
                mov     ah,0            ;keyboard input function
                int     16h

                cmp     ah,4bh          ;is left arrow ?
                jnz     no_left         ;no

                cmp     curx,0          ;yes,left moving func
                jnz     l1
                cmp     cury,0
                jnz     l3
                jmp     input
l1:             dec     curx
                jmp     l5
l3:             dec     cury
                mov     curx,79
l5:             postc
                jmp     input           ;receive next char

no_left:
                cmp     ah,4dh          ;is right arrow ?
                jnz     no_right        ;no

                cmp     curx,79         ;yes,right moving func
                jnz     r1
                cmp     cury,24
```

```
         jnz     r3
         jmp     input
r1:      inc     curx
         jmp     r5
r3:      inc     cury
         mov     curx,0
r5:      postc
         jmp     input       ;receive next char

no_right:
         cmp     ah,48h      ;is up arrow ?
         jnz     no_up       ;no

         cmp     cury,0      ;yes, up moving func
         jz      up1
         dec     cury
         postc
up1:     jmp     input       ;recrive next char

no_up:
         cmp     ah,50h      ;is down arrow ?
         jnz     no_down     ;no

         cmp     cury,24     ;yes,down moving func
         jz      d1
         inc     cury
         postc
d1:      jmp     input       ;receive next char

no_down:
         cmp     ah,4fh      ;is end key ?
         jz      setxy       ;yes,cursor moving end
         cmp     ah,01       ;is escape ?
         jnz     disp        ;no,display char

         ret                 ;yes,return to DOS

disp:    mov     ah,0ah      ;display a character
         mov     bh,0        ;video page
         mov     cx,1        ;count of repeat
         int     10h         ;video ROM call
         inc     curx        ;current column + 1
         mov     al,curx
         cmp     al,rdx      ;boundary control
         jl      next
         scroll  1           ;scroll one line
         mov     al,lux
         mov     curx,al     ;current column=lux
next:    postc
         jmp     input       ;entrn next key
```

```
setxy:
        mov     al,cury         ;current position
        mov     bl,curx         ;  of cursor
        cmp     al,luy          ;>= upper_left row# ?
        jl      s1              ;no
        cmp     bl,lux          ;>= upper_left col# ?
        jl      s2              ;no
        mov     rdy,al          ;it is new coordinate
        mov     rdx,bl          ;  of upper_left
        jmp     new             ;make a new window
s1:
        cmp     bl,rdx          ;> lower_right row # ?
        jle     s3              ;no
        jmp     input
s2:
        cmp     al,rdy          ;>= right_down col # ?
        jg      s5              ;no
s3:
        mov     luy,al          ;it is new coordinate
        mov     lux,bl          ;  oflower_right
new:
        mov     al,rdy
        sub     al,luy
        inc     al
        mov     cont,al         ;row num of new window
        clear                   ;clear screen
        scroll  cont            ;   generate new window
s5:     jmp     input
main    endp
;————————————————————————————————————————
cseg    ends                    ;end of segment
        end     main            ;end assembly
```

图 3.5　例 3.4 的程序清单

  编写这个程序时，因多次要用到清屏、生成窗口、光标定位、卷屏、光标定位等功能，因此采用宏汇编技术，把这些功能分别定义成宏指令 clear、scroll、postc 等，这样，在编写程序时，这些定义后的宏指令就和其它汇编语言指令一样直接写在程序中。

  在前面例 3.3 中，清除屏幕用的是存储器映射法，即连续往显存各单元中写入 0。本例中清除屏幕使用 BIOS 10H 的卷屏功能 ah=6 或 ah=7，给定左上角参数为(0,0)，右下角参数为(24,79)，属性为 07，然后调用 10H ROM 例行程序，则完成全屏幕清除的工作。清除屏幕的方法还有很多，在例 3.5 中，用 INT 16H 的功能 0，重置屏幕显示方式也能达到清除屏幕的效果。

  键盘输入使用 INT 16H 的功能 0，如键入一个光标移动键，则控制光标在屏幕 25×80 的范围内移动；如键入 END 键(扫描码为 4FH)，则以当前光标位置作为新设定的坐标生成新的窗口；如键入 Esc 键，则退出本程序返回 DOS；若键入其他可显示字符，则可在窗口范围内显示，若显示字符超过窗口的右边界，窗口则上卷一行，字符又从下面一行开始显示。

  光标定位后，在生成新窗口之前，有一段实现坐标代换的程序(setxy)，它把光标的行号和列号或者代换为窗口的左上角坐标，或者代换为窗口的右下角坐标，这要根据光标在原窗口的位置而定。如图 3.6,若光标处在原窗口 ABCD 的右下部分(G'AD'),则当前光标位置取代右下角坐标,并以这个新的右下角坐标和原左上角坐标作为参数产生一个新的窗口。如果光标处在原窗口的左上部分(H'ABC'),则当前光标位置取代左上角坐标,并且保持右下角坐标不变形成一个新的窗口。这种窗口的变化可进行多次,既可在原窗口的基础上扩大,也可缩小,并且任意形成正方形、长方形等各种窗口。

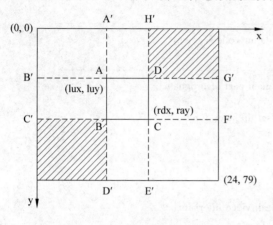

图 3.6 屏幕窗口区域

从图 3.6 中可以看出,光标如果定位在 H'DG'区域或 C'BD'区域(图中阴影部分),则该点既不能取代窗口的左上角坐标,也不能取代右下角坐标,所以这两个阴影部分属于不可代换区域,如果光标定位在这两个区域,将会产生一个很奇怪的窗口,所以程序中排除这两块区域中的光标代换。

在生成新窗口之前,还使用宏指令 clear 进行了一次清屏操作,以消除原窗口区域的显象,这在缩小窗口时尤为需要。

下面以生成一个上半屏窗口为例,说明该程序的运行过程。

(1) C：\wdex↵　清屏并在屏幕中心生成一个小窗口,如图 3.7(a)。
(2) 按下 END 键,以光标初始位置(0,0)为左上角坐标形成新窗口,如图 3.7(b)。
(3) 按光标控制键移动光标至 12 行 79 列,如图 3.7(c)。
(4) 按下 END 键,以新的右下角坐标形成新窗口,如图 3.7(d)。

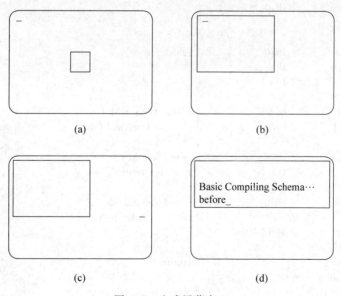

图 3.7 生成屏幕窗口

（5）在窗口内显示文本时,窗口具有边界控制和上卷功能,如图 3.7(d)。

（6）键入 ESC 键,退出程序,返回 DOS。

例 **3.5**　画横竖线程序 grid

```
TITLE GRID -- Program to draw grid on screen
;    use ROM routine
;    For 320 x 200 medium res color mode
;————————————————————————
pro_nam    segment              ;define code segment
assume     cs : pro_nam
;————————————————————————
main       proc      far        ;main part of prognam
start :
           push      ds         ;set up stack for ret
           sub       ax,ax
           push      ax

           mov       ah,0
           mov       al,4
           int       10h        ;call video interrupt

           call      hori       ;draw horizontal lines
           mov       ah,0       ;kbd input
           int       16h

           mov       ah,0       ;clear screen
           mov       al,4
           int       10h

           call      vert       ;draw vertical line
           mov       ah,0
           int       16h

           mov       ah,0       ;clear screen
           mov       al,4
           int       10h

           call      hori       ;draw horizontal
           call      vert       ;     and vertical line

           mov       ah,0       ;entre any key
           int       16h        ;     for return

           mov       ah,0       ;reset video mode
           mov       al,3       ;80 x 25 color text
           int       10h

           ret                  ;return
main       endp
;————————————————————————
hori       proc      near
;draw horizontal lines every 20 pixels
```

```
              mov     dx,0              ;row number in DX
hline:
              mov     cx,0              ;column number in CX
hdot:
              mov     al,1              ;set color to 1
              mov     ah,0ch            ;write dot function
              int     10h               ;call video ROM
              inc     cx                ;next dot
              cmp     cx,300            ;done all dots ?
              jl      hdot              ;not yet
              add     dx,20             ;next horizontal line
              cmp     dx,200            ;off the screen yet ?
              jl      hline             ;not yet
              ret
hori          endp
;————————————————————————————————————
vert          proc near
;draw one vertical lines every 20 pixels
              mov     cx,0              ;set to first line
vline:
              mov     dx,0              ;start of vert line
vdot:
              mov     al,2              ;set color to 2
              mov     ah,0ch            ;write dot funtion
              int     10h               ;call video ROM
              inc     dx                ;next dot
              cmp     dx,180            ;done all dots ?
              jl      vdot              ;not yet
              add     cx,20             ;next vertical line
              cmp     cx,320            ;off the screen yet ?
              jl      vline             ;do next line
              ret                       ;return
vert          endp
;————————————————————————————————————
pro_nam       ends                      ;end of code segment
              end     start
```

图 3.8 例 3.5 的程序清单

这是一个非常简单的绘制水平线和垂直线的程序。在这个画线的程序中使用了 BIOS 写像素功能(AH=0CH)和设置图形方式的功能(AH=0)。

汇编、连接并运行这个程序,在键盘的交互作用下,屏幕上先后画出了等距离的横线、竖线以及横竖交叉的栅栏线图案。

阅读整个程序,它完成的功能如下:

(1) 用 BIOS 显示功能调用 INT 10H,将显示方式设置为 320×200 中分辨率彩色图形方式,同时重置屏幕方式也达到了清屏的效果。

(2) 调用 hori 子程序,屏幕上从 0 行 0 列开始画一条水平线。其方法是行号不变,每写一个点,列号加 1,一直逐点画线到 300 列。然后列号重置为 0,行号递增 20,又画一条水平线,一直到屏底共画出 10 条间距为 20 的水平线。

(3) 当键盘上按动任意一个键后,执行清屏操作,调用 vert 子程序。vert 子程序完成画垂直线的工作,其方法是从 0 行 0 列开始,每写一个点,行号加 1,列号不变,一直到行号变为 180 行为止。然后重置行号为 0,列号递增 20,再开始画另一条垂直线,一共在屏幕上画出 16 条垂直线。

(4) 连续调用 hori 子程序和 vert 子程序,则屏幕上画出横竖交叉的栅栏线。当再按动一个键后,程序恢复 80×25 彩色文本显示方式,然后退出程序,返回 DOS。

图 3.9 是画线程序在屏幕上画出的横线(绿色)、竖线(红色)以及横竖交叉的栅栏线。

 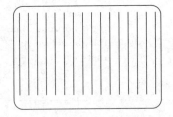 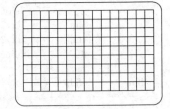

图 3.9 屏幕上的横竖线

二、实验题

实验 3.3*　字符图形程序

1. 题目:字符图形程序

2. 实验要求:制作一个图形元素表 graphic element menu,表中将所需的各种字符图形编上号码,并用存储器映射法将其显示在屏幕的左上部分,如图 3.10。

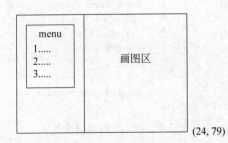

图 3.10 屏幕分区显示

将光标定位在屏幕的右半部分画图区。

上、下、左、右光标控制键控制光标在四个方向上移动,不显示其轨迹,同时要控制光标位置不超出画图区的边界。

当按动某一数字键时,在光标位置处显示出相应编号的字符图形,然后移动光标,再显示一个字符图形,最后绘制出一幅由字符组成的图形。

按动 Esc 键,退出程序,返回 DOS。

3. 提示

(1) 显示缓冲区设置在 0B800H(或 0B00H),并定义为附加段。图形元素表和其他信息定义在数据段,如:

```
menu    db    '----graph   element----'
        db    '    1    block    (odbh')
        db    '    2    O        (4fh)'
        db    '    3    /        (2fh)'
mend    db    '    4    clear    (00h)'
mlen    equ   $-mend
```

（2）清除屏幕、显示图形表以及在光标位置显示字符图形，均采用"存储器映射法"。如使用彩色显示器，在显示图形表或显示图形时，赋于彩色属性，以增强显示效果。

（3）调用 BIOS INT 10H 设置光标的功能指示光标移动的位置，以便更好地确定显示图形的位置。

（4）为了简单起见，各图形元素的属性可以固定，如按下数字键 1 后，可送给显存指定单元一个数据 02dbh，即一个绿色的方块图形。按下数字键 2 后，可送出数据 044fh，即一个红色的小圆。

（5）光标从初始位置开始，每次移动都修改其行列值，按下数字键表示光标移动结束，相应字符将送入显存，显存的地址应按照当前光标的行列位置换算出来。

（6）参考图形

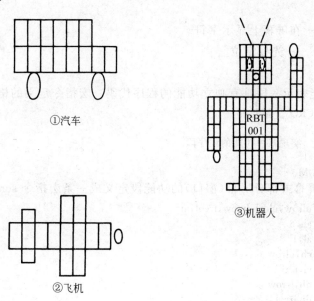

①汽车　②飞机　③机器人

图 3.11　参考图形

实验 3.4　屏幕窗口程序

1. 题目：屏幕窗口程序

2. 实验要求：在屏幕上开出三个窗口，它们的行列坐标如图 3.12 所示。

光标首先定位在右窗口最下面一行的行首(15,50)，如从键盘输入字符，则显示在右窗口，同时也显示在下窗口的最下面一行。若需要将字符显示于左窗口，则先按下←键，接着再从键盘输入字符，字符就会从左窗口的最下行开始显示，同时下窗口也显示出左窗口的内容。如若再按下→键，输入字符就会接在先前输入的字符之后显示出来。当一行

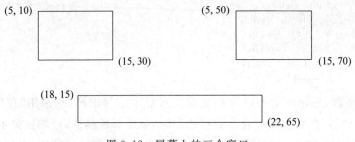

图 3.12 屏幕上的三个窗口

字符显示满后(左右窗口一行显示 20 个字符,下窗口一行显示 50 个字符),窗口自动向上卷动一行,输入字符继续显示于最低一行,窗口最高一行向上卷动后消失。

编写屏幕窗口程序时,要求将一些功能程序段定义成宏指令,如:

① scroll——向上卷动

② clear——清屏

③ get_char——接收输入字符,并判断是否是 ESCape、←或→键,然后转到相应的入口处理。

④ display——在屏幕上显示字符

⑤ pos_curse——光标定位

3. 提 示

(1) 宏指令能表示一段具有独立功能的程序代码。宏指令定义的格式如下:

宏指令名　MACRO　[哑元表]

　　…　}实现某种功能的程序段

　　ENDM

如把上卷全屏幕或部分屏幕(窗口)的功能段定义成一条宏指令 scroll,其形式为:

```
scroll  macro  ulrow,ulcol,lrrow,lrcol,att
        mov    ah,6
        mov    al,1
        mov    ch,ulrow
        mov    cl,ulcol
        mov    dh,lrrow
        mov    dl,lrcol
        mov    bh,att
        int    10h
        endm
```

如此定义之后,在程序中可直接引用宏指令 scroll,同时可赋予不同的参数而对不同的窗口进行上卷操作。如程序中需对右窗口上卷一行时调用宏指令:

　　　　scroll　5,50,15,70,70h

有关宏汇编的详细内容,请参阅教材第七章"高级汇编语言技术"。

(2) 在数据段中设置 6 个变量 lx,ly,rx,ry,dwx 和 dwy,它们分别用来记录左窗口、右窗口和下窗口的当前光标位置,以保证在返回本窗口显示字符时,能够接着前一次显示的字符串之后继续显示。

(3) 在窗口中每显示一个字符,都要修改光标的列变量,同时判断是否超出本窗口的边界,如没有超出可继续接收并显示字符,如已超出边界,则需上卷一行,同时把光标重新定位在本窗口底行的行首。

(4) 一个字符无论在左窗口显示还是在右窗口显示,它同时还要在下窗口中显示,因此要注意保存显示字符。

实验 3.5* 画栅栏线程序

1. 题目:画栅栏线程序

2. 实验要求:用图形方式在屏幕上画出横竖交叉的栅栏线,要求程序能按用户输入的起点、终点及其行距、列距画出任意条横竖线。

所需参数在显示提示信息之后,以十进制形式由键盘输入,敲入回车键表示参数输入完毕。

```
Enter  starting  point  x :  10↙
Enter  starting  point  y :  20↙
Enter  ending    point  x :  300↙
Enter  ending    point  y :  200↙
Enter  column    distance :  20↙
Enter  row       distance :  10↙
```

3. 提示

(1) 以图形方式画水平线或垂直线,就是按行列坐标在 320×200 的点阵屏幕上逐点写像素来实现的:

```
        mov  dx,   行坐标(0~199)
        mov  cx,   列坐标(0~319)
        mov  al,   像素值(0~3)
        mov  ah,   0ch        ;写像素功能
        int  10h
```

(2) 由键盘输入的十进制数据,可用一个子程序完成十进制转化为二进制的工作。

(3) 转化为二进制的 6 个数据应保存在数据段的相应单元中,以备画线时使用这些数据。

(4) 在需要重置显示方式的情况下,清屏功能应优先采用设置显示方式的方法。在显示提示信息之前,设置彩色文本方式并清屏:

```
        mov  ah,0
        mov  al,3
        int  10h
```

在画图之前,设置彩色图形方式并清屏:

```
        mov  ah,0
        mov  al,4
        int  10h
```

(5) 按动 Esc 键,则退出程序返回 DOS。在此之前,应使屏幕显示方式还原为文本方式。

3.3 键盘输入程序设计

在 PC 中,对键盘的管理是通过中断机构和 8255 可编程序外围接口芯片来实现的。在 8255 芯片中有两个端口 PA(60H)和 PB(61H),在这个硬件接口的基础上,系统在

BIOS中已配备了键盘的服务功能,因此用户可调用键盘的DOS和BIOS功能编程,也可直接在硬件接口的基础上编程。

在硬件接口的基础上编写键盘输入程序,必须了解键盘接口的信息,下图为有关端口的详细信息:

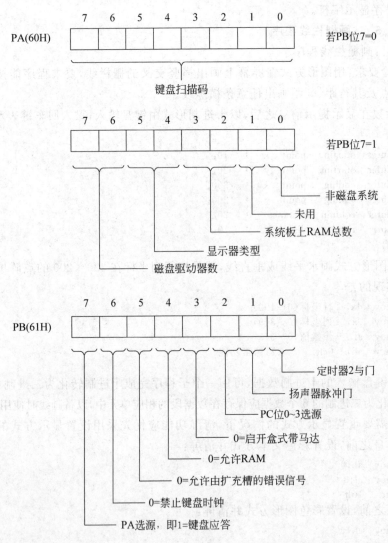

在键盘内部,有一个微处理器INTEL 8048,该处理器从系统板接收时钟信号,并读取每个键入的字符,将其扫描码放在8255外围接口芯片的PA端口(60H)内。PB端口(61H)的第6位控制着键盘的时钟信号,当键盘正常工作时,第6位应总是1,否则将锁闭键盘。PB端口的第7位置1时,可发送一个应答信号给键盘微处理器。每当按下键或放开键,在8048将其扫描码(通码或断码)送入PA端口的同时,还产生一个类型为09H的中断,该中断的任务是:①读扫描码并把应答信号送到键盘;②把扫描码转换成字符码或变换键状态;③在键盘缓冲区内设置键的字符码。用户自编键盘中断程序,可重新定义键盘上的任何键。下面是CPU读取扫描码并发出应答信号的程序段:

```
        in      al, 60H         ;从 PA 口读扫描码
        push    ax
        in      al, 61h         ;读 PB 口信息
        or      al, 80h
        out     61h, al         ;置键盘应答位
        and     al, 7fh
        out     61h, al         ;复位键盘应答位
```

示例 3.6 是直接基于键盘硬件接口基础之上进行程序设计的一个实例。

在 BIOS 中系统已开发了一些键盘 I/O 功能,用户可直接调用这些功能获取键盘的信息。从示例 3.7 和 3.8 可以体会到用户利用 BIOS 功能调用编写程序比直接在硬件基础上编写程序要方便简单得多。

一、示例

例 3.6 键盘处理演示程序 kbdio

```
TITLE kbdio.asm --- Keyboard I/O support program
;————————————————————————————————
stack   segment para    stack 'stack'
        db      256 dup(0)
stack   ends
;————————————————————————————————
data    segment para    public 'data'
buffer  db      16h dup (0)
bufpt1  dw      0
bufpt2  dw      0
; bufpt1 = bufpt2 , the buffer is empty
kbflag  db      0
prompt  db      '--- kbd_io program begin ---',0dh,0ah,'$'
scantab db 0,0,'1234567890-=',8,0
        db 'qwertyuiop[ ]',0dh,0
        db 'asdfghjkl;',0,0,0,0
        db 'zxcvbnm,./',0,0,0
        db ' ',0,0,0,0,0,0,0,0,0,0,0,0
        db '789-456+1230.'
even
oldcs9  dw ?
oldip9  dw ?
data    ends
;————————————————————————————————
code    segment para    public          'code'
start   proc    far
        assume  cs:code, ds:data
        push    ds                      ;save for return
        mov     ax,0
        push    ax
        mov     ax,data                 ;set DS to data seg
        mov     ds,ax

        cli
        mov     al,09                   ;save interrupt vector
```

```
              mov       ah,35h              ;of KBD BIOS routing
              int       21h
              mov       oldcs9,es
              mov       oldip9,bx

              push      ds                  ;set interrupt vector
              mov       dx,offset kbint     ;   of kbint
              mov       ax,seg kbint
              mov       ds,ax
              mov       al,09
              mov       ah,25h
              int       21h
              pop       ds

              in        al,21h              ;set kbd interrupt
              and       al,0fdh             ;   mask bit
              out       21h,al

              mov       dx,offset prompt
              mov       ah,9
              int       21h
              sti
forever:
              call      kbget               ;wait enter a key
              test      kbflag,80h
              jnz       endint
              push      ax
              call      dispchar            ;display the character
              pop       ax
              cmp       al,0dh
              jnz       forever
              mov       al,0ah
              call      dispchar            ;display CR/LF
              jmp       forever             ;loop for continue
endint:
              mov       dx,oldip9           ; interrupt vector
              mov       ax,oldcs9
              mov       ds,ax               ;restore old
              mov       al,09h
              mov       ah,25h
              int       21h

              ret
start         endp
;------------------------------------------------------------
kbget         proc      near
              push      bx
              cli                           ;interrupt back off
              mov       bx,bufpt1           ;get pointer to head
              cmp       bx,bufpt2           ;test empty of buffer
              jnz       kbget2              ;no,fetch a character
              cmp       kbflag,0
              jnz       kbget3
```

```
                sti                             ;allow an interrupt to occur
                pop     bx
                jmp     kbget                   ;loop until something in buf
        kbget2:
                mov     al,[buffer+bx]          ;get ascii code
                inc     bx                      ;inc a buffer pointer
                cmp     bx,16                   ;at end of buffer ?
                jc      kbget3                  ;no,continue
                mov     bx,0                    ;reset to buf beginning
        kbget3:
                mov     bufpt1,bx               ;store value in variable
                pop     bx
                ret
        kbget   endp
;————————————————————————————————————————————————————————————————
        kbint   proc    far                     ;keyboard interrupt routine
                push    bx
                push    ax

                in      al,60h                  ;read in the character
                push    ax                      ;save it
                in      al,61h                  ;get the control port
                or      al,80h                  ;set acknowledge bit for kbd
                out     61h,al
                and     al,7fh                  ;reset acknowledge bit
                out     61h,al

                pop     ax;                     recover scan code
                test    al,80h                  ;is press or release code?
                jnz     kbint2                  ;is release code,return
                mov     bx,offset scantab
                xlat    scantab                 ;ascii code to AL
                cmp     al,0
                jnz     kbint4
                mov     kbflag,80h
                jmp     kbint2
        kbint4:
                mov     bx,bufpt2               ;buffer tail pointer
                mov     [buffer+bx],al          ;ASCII fill in buffer
                inc     bx
                cmp     bx,16                   ;is end of buffer?
                jc      kbint3                  ;no
                mov     bx,0                    ;reset to buf beginning
        kbint3:
                cmp     bx,bufpt1               ;is buffer full?
                jz      kbint2                  ;yes, lose character
                mov     bufpt2,bx               ;save buf tail pointer
        kbint2: cli
                mov     al,20h                  ;end of interrupt
                out     20h,al
                pop     ax
                pop     bx
                sti
```

```
                iret                            ;interrupt return
    kbint       endp
;──────────────────────────────────────────────
    dispchar    proc    near                    ;(AL)=displaying char.
                push    bx
                mov     bx,0
                mov     ah,0eh
                int     10h                     ;call video routine
                pop     bx
                ret
    dispchar    endp
;──────────────────────────────────────────────
    code        ends                            ;end of code segment
                end     start
```

图 3.13 例 3.6 的程序清单

例 3.6 是一个完整的键盘处理程序,不过它对按键的处理做了一些简化。在这个演示程序中,要求完成对键盘的中断检测,并把来自键盘的 83 个键的扫描码转换成相应的 ASCII 字符码。

整个程序由下面几部分组成:

主程序 start:键盘中断向量的保存、设置与恢复;设置中断屏蔽位并开中断;从缓冲区中读取键入字符并显示在屏幕上。

子程序 kbget:检测并等待键盘中断,如有键盘输入,则从缓冲区中取出字符并进行队列管理。

键盘中断处理程序 kbint:从输入口 PA(60H)读入按键的扫描码并返回应答信号;对通码进行转换,将转换后的 ASCII 码存入队列尾。

子程序 dispchar:调用 BIOS 显示功能(int 10h)显示键入的字符。

键盘是通过中断方式工作的,由于键盘中断的出现是完全随机的,因此要求键盘 I/O 程序要缓冲或保留它接收的任一键盘输入。为此数据段中定义了一个"先进先出"缓冲区 (buffer),也叫循环队列。在这个队列中存入字符和读取字符分别由两个指针 bufpt2(尾指针)和 bufpt1(头指针)指示,当 bufpt1=bufpt2 时,说明队列中没有数据,程序循环等待下一次键盘中断输入。当指针增量超过队尾(buffer+16)时,头尾指针可循环指向队列的开头(buffer+0),从而保证总是从队列中以存放时的顺序收回数据。如果队列存满时接收了字符,则就简单地忽略这个字符。

下面是循环队列的几种工作状态:

① 循环队列空状态(bufpt1=bufpt2)

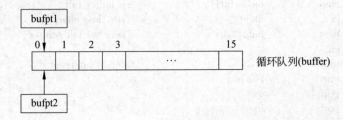

② 键盘输入的字符进入队列

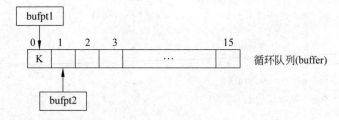

③ bufpt1≠bufpt2,则从队列取出一个字符

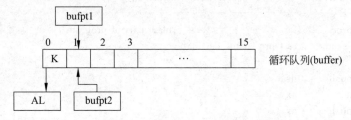

④ 指针值增量超过队尾时,指针循环指向队列开头

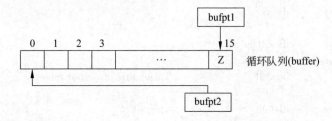

为了把接收的扫描码转换成相应的 ASCII 码,数据段中还定义了一个转换表 scan-tab,该转换表是一个简化了的字符码表,除可显示字符外,其他控制键、功能键都作无效键处理,其字符码 0 也不进入队列。当在键盘上按动任一控制键或功能键时,程序将标志字节 kbflag 置为 80h,以控制结束键盘输入,返回 DOS。

例 3.7 键盘输入程序 keyboard

这个程序利用中断类型 16H 调用键盘 I/O ROM 例行程序并显示出由例行程序送回的扫描码和 ASCII 码。显示的格式如下:

```
SCAN   ASCII   CHAR
1E     61      a
30     62      b
2E     63      c
```

下面是键盘输入程序的打印清单:

```
TITLE KEYBOARD --- Keyboard I/O test
;   and prints out scan code and ASCII of any key
;————————————————————————————————
display   equ   2h      ;display character fuc
doscall   equ   21h     ;DOS interruot routine
```

```
;------------------------------------------------
        data    segment
        mess    db      0dh,0ah,'            SCAN   ASCII   CHAR$'
        mess1   db      '   $'
        data    ends
;------------------------------------------------
        stack   segment para stack 'stack'
                db 64 dup (?)
        stack   ends
;* * * * * * * * * * * * * * * * * * * * * * * * * * * * *
        pro_nam segment                 ;define code segment
;------------------------------------------------
        main    proc    far             ;main part of program
                assume  cs:pro_nam, ds:data
        start:
                push    ds
                xor     ax,ax
                push    ax

                mov     ax,data
                mov     ds,ax

                mov     dx,offset mess
                mov     ah,9h
                int     doscall         ;print title of output table

                mov     dl,0dh          ;print CR
                mov     ah,display
                int     doscall

                mov     dl,0ah          ;print LF
                mov     ah,display
                int     21h
        again:
                mov     ah,0            ;read character funct
                int     16h             ;keyboard I/O ROM call
                cmp     al,03h          ;is ctrl_C ?
                jz      exit            ;yes,exit

                mov     bx,ax           ;move scan & char to BX
                call    binihex         ;print scancode & char
                mov     dx,offset mess1 ;print space
                mov     ah,9h
                int     doscall

                mov     dl,bl           ;print character
                mov     ah,display      ;   in ASCII
                int     doscall
                mov     dl,0dh          ;print return
```

```
              mov       ah,display
              int       doscall
              mov       dl,0ah              ;print linefeed
              mov       ah,display
              int       doscall

              jmp       again               ;get another one
exit:         ret                           ;return form program to DOS
main          endp                          ;end of main part of program
;------------------------------------------------------------
binihex       proc      near
;subroutine to convert binnary in BX
;    to hex on console screen
              mov       ch,4                ;number of digits
rot1:
              push      ax
              mov       dx,offset mess1     ;print 5 blanks between
              mov       ah,9h               ;   each two columns
              int       doscall
              pop       ax
rotate:       mov       cl,4                ;set count to 4 bits
              rol       bx,cl               ;left digit to right
              mov       al,bl               ;move to AL
              and       al,0fh              ;mask off left digit
              add       al,30h              ;convert hex to ASCII
              cmp       al,3ah              ;is it > 9?
              jl        printit             ;no, so 0 to 9 digit
              add       al,7h               ;yes, so A to F digit
printit:      mov       dl,al               ;put ASCII char in DL
              mov       ah,display          ;display output funct.
              int       doscall             ;call DOS
              dec       ch
              cmp       ch,2                ;done 2 digits?
              jg        rotate              ;not yet
              jz        rot1
              cmp       ch,0                ;done 4 digits?
              jnz       rotate
              ret                           ;done subroutine
binihex       endp
;------------------------------------------------------------
pro_nam       ends                          ;end of code segment
;* * * * * * * * * * * * * * * * * * * * * * * * * * * * * *
              end       start               ;end assembly
```

图 3.14 例 3.7 的程序清单

这个程序比较简单,其中心部分是使用了 INT 16H 键盘功能调用并调用一个显示十六进制数据的子程序:

```
         mov    ah,0       ;读字符功能
         int    16h        ;BIOS 键盘调用
         mov    bx,ax
```

```
            call    binihex              ;打印扫描码和 ASCII 码
```

因为 BIOS 例行程序回送的扫描码和 ASCII 码放在 AH 和 AL 中,因此可通过打印出 AX 的内容查看按动每个键的情形。运行这个程序,可使我们对键盘上每个键的扫描码、ASCII 码及其显示字符(如果存在可显示的形式)之间的对应关系一目了然。弄清键盘的有关内容,对于编写游戏程序和那些使用功能键或光标控制键的程序都是很重要的,例 3.8 字处理程序就是利用扫描码进行功能控制的程序实例。

例 3.8 字处理演示程序 wspp

```
TITLE WSPP --- Program of word process function
;       for insert, left and right
;――――――――――――――――――――――――――――――
dseg        segment                      ;define data segment
   kbd_buf  db      96    dup(' ')       ;input buffer
   cntl     db      16    dup(0)         ;char number of rows
   bufpt    dw      0                    ;buffer head pointer
   buftl    dw      0                    ;buffer tail pointer
   colpt    db      0                    ;current col pointer
   rowpt    db      0                    ;current row pointer
   rowmx    dw      0                    ;maxium row number
dseg        ends
;――――――――――――――――――――――――――――――
curs        macro   row,col              ;position cursor macro
            mov     dh,row
            mov     dl,col
            mov     bh,0
            mov     ah,2
            int     10h
            endm
;――――――――――――――――――――――――――――――
cseg        segment                      ;define code segment
main        proc    far
            assume  cs:cseg, ds:dseg, es:dseg
start:
            push    ds                   ;save for retuen to DOS
            sub     ax,ax
            push    ax
            mov     ax,dseg              ;dseg addr into ds,es
            mov     ds,ax
            mov     es,ax

            mov     buftl,0              ;initialize pointers
            mov     colpt,0
            mov     rowpt,0
            mov     bufpt,0
            mov     rowmx,0
            mov     cx,length cntl       ;initialize cntl area
            mov     al,0
            lea     di,cntl
            cld
            rep     stosb
```

```
              mov     ah,6                    ;clear screen
              mov     al,0
              mov     cx,0
              mov     dh,24
              mov     dl,79
              mov     bh,07
              int     10h
              curs    0,0                     ;place cursor at (0,0)
read_k :
              mov     ah,0                    ;read char from kbd
              int     16h                     ;call ROM routine
              cmp     al,1bh                  ;is escape ?
              jnz     arrow
              ret                             ;yes, return to DOS
arrow :
              cmp     ah,4bh                  ;is left arrow ?
              jz      left                    ;yes, moving cursor
              cmp     ah,4dh                  ;is right arrow ?
              jz      right                   ;yes
;-----------------------------------------------------------
inst :        jmp     ins_k
left :        jmp     left_k
right :       jmp     right_k
;-----------------------------------------------------------
ins_k :                                       ;insert a character
              mov     bx,bufpt
              mov     cx,buftl
              cmp     bx,cx                   ;bufpt = buftl ?
              je      km                      ;yes,char into buffer
              lea     di,kbd_buf              ;no, buffer move
              add     di,cx                   ;   a byte backward
              mov     si,di
              dec     si
              sub     cx,bx
              std
              rep     movsb
km :
              mov     kbd_buf[bx],al          ;char into buffer
              inc     bufpt                   ;inc head pointer
              inc     buftl                   ;inc tail pointer
              cmp     al,0dh                  ;insert a CR ?
              jnz     kn                      ;no
              lea     si,cntl                 ;yes,move the count
              add     si,rowmx                ;   of each row
              inc     si                      ;      backward
              mov     di,si
              inc     di
              mov     cx,rowmx
              sub     cl,rowpt
              std
              rep     movsb

              mov     bl,rowpt                ;adjust the counts
```

```
        xor     bh,bh               ;   of current row
        mov     cl,colpt            ;       and next row
        mov     ch,cntl[bx]
        sub     ch,colpt
        mov     cntl[bx],cl
        mov     cntl[bx+1],ch

        mov     ax,rowmx            ;clear displaying row
        mov     bh,07               ;   use scroll function
        mov     ch,rowpt
        mov     dh,24
        mov     cl,0
        mov     dl,79
        mov     ah,6
        int     10h

        inc     rowpt               ;point to next row
        inc     rowmx               ;inc max row count
        mov     colpt,0             ;point to 0 column
        jmp     short kp
kn:
        mov     bl,rowpt
        xor     bh,bh
        inc     cntl[bx]            ;inc current row count
        inc     colpt               ;point to next column
kp:
        call    dispbf              ;display input buffer
        curs    rowpt,colpt         ;position the cursor
        jmp     read_k

left_k:
        cmp     colpt,0             ;is at 0 column ?
        jnz     k2                  ; no
        cmp     rowpt,0             ;is at 0 row ?
        jz      lret                ; yes,cursor is unmove
        dec     rowpt               ;point to upper row
        mov     al,rowpt
        lea     bx,cntl
        xlat    cntl
        mov     colpt,al            ;point to tail of row
        jmp     k3
k2:     dec     colpt               ;dec column pointer
k3:     dec     bufpt               ; dec buffer point
        curs    rowpt,colpt         ;position cursor
lret:   jmp     read_k
right_k:
        mov     bx,bufpt            ;is at tail of file ?
        cmp     bx,buftl
        je      rret                ;yes,cursor unmoved
        inc     colpt               ;point to next column
        cmp     kbd_buf[bx],0dh     ;is CR ?
        jnz     k4                  ;no
        inc     rowpt               ;yes,point to next row
```

```
            mov     colpt,0             ; and 0 column
   k4:      inc     bufpt               ; adjust buffer pointer
            curs    rowpt,colpt         ; position cursor
   rret:    jmp     read_k
   ;--------------------------------------------------------------
   dispbf   proc    near                ; display char of buffer
            mov     bx,0
            mov     cx,96
            curs    0,0
   disp:    mov     al,kbd_buf[bx]
            push    bx
            mov     bx,0700
            mov     ah,0eh
            int     10h                 ; call ROM routine
            pop     bx
            cmp     al,0dh              ; is CR ?
            jnz     kk
            mov     al,0ah              ; yes,display LF
            mov     ah,0eh
            int     10h                 ; video call
   kk:      inc     bx
            loop    disp
            ret
   dispbf   endp
   ;--------------------------------------------------------------
   main     endp                        ; end main part of program
   ;--------------------------------------------------------------
   cseg     ends
            end     start               ; end assembly
```

图 3.15 例 3.8 的程序清单

编写一个功能完整的实用文字处理软件是一件比较复杂的工作。类似 wordstar、pced 等字处理软件,使用功能键、光标控制键等作为命令键完成了许多文字编辑工作,包括光标移动,字符插入、删除、查找,文件的存储、复制、删除、打印等功能。

本示例程序是一个只具有左右移和字符插入功能的简化程序。

程序采用 case 分支结构,根据输入键的字符码和扫描码,由一段 JMP 指令组成跳转程序,分别转入不同的处理程序段。从例 3.4 的运行结果中知道,左箭头键的扫描码是 4bh,右箭头键的扫描码是 4dh。在按动键时,只要接收的是可显示字符(ASCII 码不为 0),就作为输入字符存入文件尾或插入文件中。

在这些功能段中,最重要的工作是管理当前光标位置的指针和键盘缓冲区的头、尾指针,为此,在数据段中,设置了一个 60H 字节的缓冲区(kbd_buff),两个字的缓冲头、尾指针(bufpt & buftl),指示光标位置的行、列指针(rowpt & colpt),根据需要还设置了 10H 行的字符计数单元和一个字的最大输入行的记录单元。如此设置的缓冲区及其他指针单元,在调试程序时,可正好用 DEBUG 的 D 命令将全部信息一次显示出来。程序调试好后,可根据实际扩充数据区。

图 3.16 是字符插入,光标左移和光标右移三个功能段的程序框图。

在插入字符的功能段(ins_k)中,先比较头尾指针以区别是在文件尾输入字符,还是

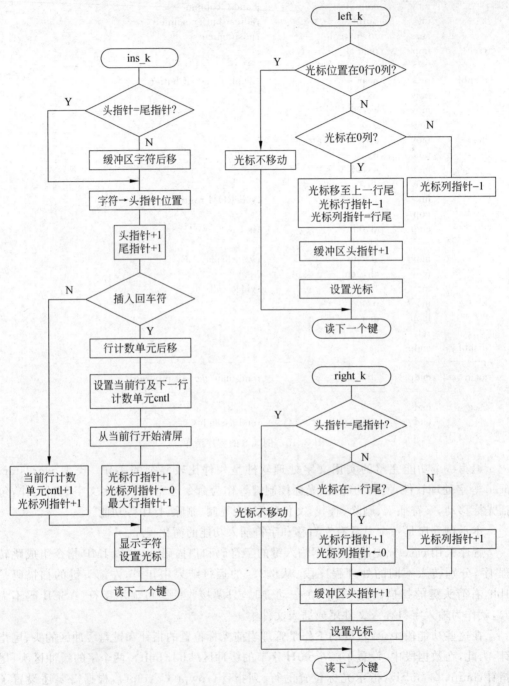

图 3.16 例 3.8 的程序框图

在文件中光标指示的位置插入字符。如果是输入字符,则将字符存入缓冲区,然后修改光标指针,如果是在文件中插入一个字符,则缓冲区的字符均要后移一个字节,以插入输入的字符。

插入的字符如果是回车符(0DH),还可能分三种情况:(1)重新开一行输入字符;(2)插入一空行;(3)把一行拆分为两行。这时各行的计数单元在这三种情况下分别为(1)开始新一行的计数;(2)空行的计数值为0,其余行的计数值均后移一个字节;(3)列指针之前的值为当前行的计数值,列指针之后的字符数为下一行的计数值,其余行的计数值也后移一个字节。光标指针在这三种情况下均指向下一行的行首。

每当缓冲区中字符存储情况有变化(插入或删除一个字符)时,屏幕上文本的显示也要相应有变化,对此我们采用了一种控制最简单的显示方法,即每次插入(或删除)字符都把缓冲区的字符从头至尾(尾指针指示文件尾)显示一遍,当显示的字符是回车符时,紧接着再显示一个换行符(0AH)。这里要注意的是,用这种方法显示缓冲区字符,在有的情况下会出现"重映"现象。例如,先输入了四个字符:abcd,当我们要在 bc 之间插入一个回车符时,屏幕上应显示成两行,第一行为:ab,第二行为:cd,而实际上第一行显示成:abcd,第二行为:cd。这是什么原因呢?这是因为当调用 dispbf 子程序逐个显示字符时,当显示到字符 b 之后新插入的回车符,只控制光标回复到行首,而原来在本行显示的字符 cd 仍然保留在原显示位置,这个问题在实验 3.7 的删除处理中也同样存在。解决的办法是,在显示之前,先用上卷功能将原显示字符清除,再逐一显示缓冲区的字符,这样就克服了"重映"现象。具体指令如下:

```
        mov   ax,rowmx     ;(ax)=最大输入行
        mov   bh,07        ;卷入行属性
        mov   ch,rowpt     ;左上角行号为当前光标行
        mov   dh,24        ;右下角行号为 24 行
        mov   cl,0         ;左上角列号为 0
        mov   dl,79        ;右下角列号为 79
        mov   ah,6         ;上卷功能
        int   10h          ;BIOS 显示调用
```

处理左右移功能时,要特别注意的是,指示光标位置的行列指针的变化和缓冲区头指针的移动要同步,否则会使字符插入(或删除)的位置和光标的指示不对应,引起字处理的混乱。

二、实验题

实验 3.6 扩充键盘处理功能的程序

1. 题目: 扩充键盘处理功能的程序

2. 实验要求:在示例 3.6 程序的基础上,增加 left_shift 和 right_shift 键的功能,即在按下 left_shift 键或 right_shift 键的同时,又按下 0~9 或 a~z 等键,则 CPU 取得并显示的是键的上档符号或大写字母。

3. 提示:

(1) left_shift 键的扫描码为 42d,right_shift 键的扫描码为 54d。

(2) 根据读取的扫描码可判别是否按下了 left_shift 键或 right_shift 键,并把按动的状态记录在一个标志单元(如 kbflag)中。设按下 right_shift 键,kbflag 的第 0 位置 1,按下 left_flag 键,kbflag 的第 1 位置 1,放开左或右 shift 键,kbflag 恢复为 0。

(3) 当 kbflag 的第 0 位或第 1 位为 1 时,再按下数字键或字母键,则应被转换为上挡

符号或大写字母,为此应再设置一个字符转换表 shiftab。当未按动 Shift 键时,所有按键通过 scantab 转换(同例 3.6);当按下 Shift 键时,按键应通过有相应上档符及大写字母的 shiftab 转换表。上挡符及大写字母的转换表如下:

```
shiftab   db   0,0,'! @#$%^&*()—+',0,0
          db   'QWERTYUIOP{}',0dh,0
          db   'ASDFGHJKL:"',0,0,0
          db   'ZXCVBNM<>? ',0,0,0
          db   ' ',26  dup(0)
```

(4) 按动除 Shift 键以外的控制键或功能键时,退出程序返回 DOS。

实验 3.7* 扩充字处理功能的程序

1. 题目:扩充字处理功能的程序
2. 实验要求:在示范的例 3.8 字处理程序的基础上,再增加光标上移、光标下移以及删除字符的功能。

当按动上箭头键(扫描码为 48h)时,光标移到上一行。如果上一行的字符数小于当前光标所在位置的列号,则光标移至上一行行尾,其它情况光标垂直上移;如果光标已在文件的首行,则按动上箭头键不能引起光标移动。

当按动下箭头键(扫描码为 50h)时,光标移到下一行。如果下一行字符数小于当前光标所在位置的列号,则移至下一行行尾,否则光标垂直下移至下一行。如果光标位置已在文件的最后一行,则按动下箭头键不能引起光标移动。

当按下删除键 DEL(扫描码为 53h)时,光标左边的一个字符被删除,本行后续字符均前移一个位置。如果删除的字符是回车符(0dh),则将下一行连接到上一行行尾,以下各行均上移一行,也就是整个文件的总行数少 1。

3. 提示

(1) 光标垂直上下移时,列指针不变,行指针±1。

(2) 如果光标要移至上一行尾或下一行尾时,可使移动后的光标列指针 colpt 等于上一行或下一行的字符计数值 cntl。

(3) 缓冲区的头指针 bufpt 所指示的字符和光标所指示的字符始终要一致,它们的移动也要保持同步。例如,光标垂直上移时,行指针减 1,列指针不变,头指针 bufpt 的修改值为:(当前头指针值)-(上一行字符计数值)-1。光标移至上一行尾时,光标行指针-1,列指针=上一行字符计数值,bufpt 的修改值为:(当前头指针值)-(当前光标列号)-1

(4) 当删除一个回车符时,上下两行连接在一起成为一行,这新的一行的字符计数值 cntl 应是原来两行字符数之和,以下各行的字符计数值 cntl 均前移一个字节,最后一个计数单元补以 0 值,缓冲区的处理即用前移一字节的办法删去头指针指示的前一个字节的内容,最后一个缓冲区字节要填以"空"。

(5) 在显示缓冲区时,要注意"重映"现象,即删除一行中的某个字符时,该行的最后一个字符仍会显现。这时可在显示缓冲区字符之前,用下面的指令将有删除字符的一行清除一次:

```
    mov  al,1      ;卷动一行
    mov  bh,07     ;行属性
```

```
        mov   ch,rowpt       ;左上角行号和右下角行号
        mov   dh,ch          ;为光标所在行
        mov   cl,colpt       ;左上角列号为当前列号
        mov   dl,79          ;右下角列号为 79
        mov   ah,6           ;上卷功能
        int   10h            ;BIOS 显示功能调用
```

3.4 中断程序设计

中断是计算机自动执行专门例行程序的过程。ROM BIOS 和 DOS 中都有许多中断例行程序(又称中断处理子程序)，这些例行程序或由程序中的 INT 指令执行，或由 I/O 设备的中断机构来调用。

所有的 IBM 微机都用 INTEL 8259 可编程中断控制器管理硬件中断。8259 芯片内设有判优响应机构，除 AT 机有 16 个优先级外，其它机器都有 8 级(IRQ0～IRQ7)。8259 可编程的端口有 21H 中断屏蔽寄存器和 20H 中断命令寄存器。汇编程序通过对 21H 端口的各位分别置 0 或置 1 来控制各中断级是被允许或被禁止。如：

```
        ;允许键盘和定时器中断
        in    al,21h
        and   al,11111100b
        out   21h,al
```

编写硬件中断处理程序时，应在中断结束之前，将中断命令寄存器的中断结束位置 1，以清除当前的中断级，这用下面两行程序完成：

```
        mov   al,20h
        out   20h,al
```

另外，所有中断的执行都取决于程序状态字寄存器的中断标志位 IF，当该位为 1 时，任何中断屏蔽寄存器允许的中断请求都可实现，当该位为 0 时，所有硬件中断都不可能实现。要注意的是，当调用中断例行程序时，计算机会自动禁止硬件中断(IF＝0)，因此若在中断处理过程中，允许响应硬件中断，就应在中断处理程序中加上 STI 指令。

系统中预制的 DOS 或 BIOS 中断例行程序，给用户编制程序带来极大的方便，但在某些情况下，如系统要配制专用设备，或扩充原中断例行程序的功能，或需设备执行完全不同于原中断例行程序的功能时，就需要用户自编中断处理程序。

用户自编中断处理程序有一个基本的形式：

(1) 将原中断向量保存在堆栈中或自设的存储单元中。

```
        mov   ah,35h         ;取中断向量的功能调用
        mov   al,int_type    ;中断类型
        int   21h            ;段址放入 ES,偏址放入 BX
        mov   keep_ip,bx     ;保存偏移地址
        mov   keep_cs,es     ;保存段地址
```

(2) 设置自编处理程序的中断向量：

```
        push  ds             ;保存 DS
```

```
        mov     dx,offset routine       ;自编中断例行程序偏移地址放入 DX
        mov     ax,seg routine
        mov     ds,ax                   ;自编中断例行程序段地址放入 DS
        mov     al,int_type             ;中断类型放入 AL
        mov     ah,25h                  ;设置中断向量的功能调用
        int     21h                     ;改变中断向量
        pop     ds                      ;恢复 DS
```

注意：当功能 25H 改变中断向量时会自动禁止硬件中断，因此在设置新的中断向量时，硬件中断不会使用中断向量。

（3）当中断程序结束时，必须恢复原来的中断向量，否则后续程序不能正确使用系统提供的例行程序。

```
        push    ds                      ;保存 DS
        mov     dx,keep_ip              ;取出保存的偏移地址
        mov     ax,keep_cs              ;取出保存的段地址
        mov     ds,ax
        mov     al,int_type             ;中断类型
        mov     ah,25h                  ;设置中断向量
        int     21h                     ;DOS 调用
        pop     ds                      ;恢复 DS
```

（4）编写中断处理子程序以完成中断处理功能，结束时由 IRET 返回中断点。

```
routine proc    far
        push    ax                      ;保存所有已修改的寄存器
        sti                             ;开中断
        …                               ;中断处理功能

        …
        mov     al,20h                  ;中断结束命令
        mov     20h,al                  ;送 8259 20h 端口
        pop     ax                      ;恢复寄存器
        iret                            ;中断返回
routine endp
```

只有在编写硬件中断时，才需要向 8259 发出中断结束命令，否则会导致同级或高级中断的屏蔽，特别是键盘中断被屏蔽，会造成系统瘫痪。而在扩充现有中断功能的自编例行程序中，不需这两行程序。例如，在扩充定时器中断功能的 INT 1CH 中断例行程序中，就不应送出中断结束命令。

编写中断程序还应注意正确设置 DS，例如在设置和恢复中断向量时，要访问程序内部数据变量，也要用 DS 作为调用参数寄存器，此时应注意 DS 的设置。另外某些硬件中断例行程序在程序起始处将 DS 置成了 ROM 的数据段，所以在该中断的嵌套例行程序中，如要访问自编程序的数据变量，则必须将 DS 切换成用户数据段。

一、示例

例 3.9 打字计时程序 type_ex

这是一个在键盘上练习打字并统计时间的实用程序。键盘打字采用 09 类型的键盘中断来取得输入字符并将字符显示在屏幕上；统计时间利用 08 类型的定时器中断所嵌套

的 1CH 软中断来计时。当一个句子输入完后（回车键作为结束符），屏幕上以 min：sec：msec 的格式显示出键入字符的时间。在每次打字之前，屏幕上先显示出一个例句，然后打字员按照例句，将句中字符通过键盘输入。这个过程可反复进行，当键入一个功能键时，退出打字计时程序。

图 3.17 是打字计时程序的程序清单。

```
TITLE TYPE_EX --- Test time for typing exercise
;------------------------------------------------
stack      segment   para    stack   'stack'
           db        256             dup(0)
top        label     word
stack      ends
;------------------------------------------------
data       segment   para    public  'data'
buffer     db        16h dup (0)
bufpt1     dw        0
bufpt2     dw        0
; bufpt1 = bufpt2 , the buffer is empty
kbflag     db        0
prompt     db '   * PLEASE PRACTISE TYPING *',0dh,0ah,'$'
scantab    db 0,0,'1234567890-=',8,0
           db 'qwertyuiop[]',0dh,0
           db 'asdfghjkl;',0,0,0,0
           db 'zxcvbnm,./',0,0,0,0
           db ' ',0,0,0,0,0,0,0,0,0,0,0,0,0
           db '789-456+1230.'
even
oldcs9     dw ?
oldip9     dw ?
;------------------------------------------------
str1       db 'abcd efgh ijkl mnop qrst uvwx yz.'
           db 0dh,0ah,'$'
str2       db 'christmas is a time of joy and love.'
           db 0dh,0ah,'$'
str3       db 'store windows hold togs and gifts.'
           db 0dh,0ah,'$'
str4       db 'people send christmas cards and gifts.'
           db 0dh,0ah,'$'
str5       db 'santa wish all people peace on earth.'
crlf       db 0dh,0ah,0ah,'$'
colon      db ' : ','$'
even
saddr      dw      str1, str2, str3, str4, str5
count      dw      0
sec        dw      0
min        dw      0
hours      dw      0
save_1c    dw      2   dup (?)
data       ends
;------------------------------------------------
code       segment
```

```
        assume    cs : code, ds : data, es : data, ss : stack
        main      proc    far
        start :
                  mov     ax,stack              ;set up stack
                  mov     ss,ax
                  mov     sp,offset top

                  push    ds                    ;save ds : 0 for return
                  sub     ax,ax
                  push    ax
                  mov     ax,data               ;set DS to data segment
                  mov     ds,ax
                  mov     es,ax

                  mov     ah,35h                ;save interrupt vector
                  mov     al,09h                ;  of keyboard
                  int     21h
                  mov     oldcs9,es
                  mov     oldip9,bx

                  push    ds                    ;set interrupt vector
                  mov     dx,seg kbint          ;  of kbint
                  mov     ds,dx
                  mov     dx,offset kbint
                  mov     al,09h
                  mov     ah,25h
                  int     21h
                  pop     ds

                  mov     ah,35h                ;save interrupt vector
                  mov     al,1ch                ;  of timer
                  int     21h
                  mov     save_1c,bx
                  mov     save_1c+2,es

                  push    ds                    ;set interrupt vector
                  mov     dx,seg clint;         ;  of clint
                  mov     ds,dx
                  mov     dx,offset clint
                  mov     al,1ch
                  mov     ah,25h
                  int     21h
                  pop     ds

                  in      al,21h                ;clear kbd & timer
                  and     al,11111100b          ;   mask bit
                  out     21h,al
        first :   mov     ah,0                  ;set video mode
                  mov     al,3                  ;80 x 25 color text
                  int     10h

                  mov     dx,offset prompt
                  mov     ah,9                  ;display kbd message
```

```
                int     21h

                mov     si,0
        next:   mov     dx,saddr[si]            ;display sentence
                mov     ah,09h
                int     21h

                mov     count,0                 ;set initial value
                mov     sec,0
                mov     min,0
                mov     hours,0

                sti                             ;set IF flag
        forever:
                call    kbget                   ;wait enter a key
                test    kbflag,80h
                jnz     endint
                push    ax
                call    dispchar                ;display the character
                pop     ax
                cmp     al,0dh
                jnz     forever
                mov     al,0ah
                call    dispchar                ;display CR / LF

                call    disptime                ;display typping time

                lea     dx,crlf                 ;display CR / LF
                mov     ah,09h
                int     21h

                add     si,2                    ;update pointer
                cmp     si,5*2                  ;end of sentences ?
                jne     next                    ;no,display next
                jmp     first                   ;yes,display first

        endint: cli                             ;end of typing
                push    ds
                mov     dx,save_1c
                mov     ax,save_1c+2
                mov     ds,ax
                mov     al,1ch                  ;reset interrupt vector
                mov     ah,25h                  ;    of type 1ch
                int     21h
                pop     ds

                push    ds
                mov     dx,oldip9
                mov     ax,oldcs9
                mov     ds,ax
                mov     al,09h                  ;reset interrupt vector
                mov     ah,25h                  ;    of type 09h
                int     21h
```

· 111 ·

```
                pop     ds

                sti
                ret                             ;return to DOS
main            endp
;------------------------------------------------
clint           proc    near                    ;timer int routine
                push    ds                      ;save ROM data area
                mov     bx,data                 ;set data segment
                mov     ds,bx

                lea     bx,count
                inc     word ptr [bx]           ;increment count
                cmp     word ptr [bx],18        ;1 sec = 18 count
                jne     return
                call    inct                    ;update sec and min
adj:
                cmp     hours,12                ;update hours
                jle     return
                sub     hours,12
return:
                pop     ds
                sti
                iret                            ;interrupt return
clint           endp
;------------------------------------------------
inct            proc    near                    ;update sec and min
                mov     word ptr [bx],0
                add     bx,2
                inc     word ptr [bx]
                cmp     word ptr [bx],60
                jne     exit
                call    inct
exit:           ret                             ;return to clint
inct            endp
;------------------------------------------------
disptime        proc    near
;subroutine to display typping time for min:sec:msec
                mov     ax,min
                call    bindec                  ;display min

                mov     bx,0                    ;display ':'
                mov     al,':'
                mov     ah,0eh
                int     10h
                mov     ax,sec                  ;display sec
                call    bindec

                mov     bx,0                    ;display ':'
                mov     al,':'
                mov     ah,0eh
                int     10h
```

```
                mov     bx,count                ;count convert to ms
                mov     al,55d
                mul     bl
                call    bindec                  ;display ms

                ret                             ;return to main
disptime        endp
;------------------------------------------------
bindec          proc    near
; subroutine to convert binary in AX to decimal
                mov     cx,100d
                call    decdiv
                mov     cx,10d
                call    decdiv
                mov     cx,1
                call    decdiv
                ret                             ;return to disptime
bindec          endp
;------------------------------------------------
decdiv          proc    near
; sub- subroutine divide number in AX by CX
                mov     dx,0                    ;number hight half
                div     cx

                mov     bx,0
                add     al,30h                  ;convert to ASCII
                mov     ah,0eh                  ;display function
                int     10h

                mov     ax,dx
                ret                             ;return to bindec
decdiv          endp
;* * * * * * * * * * * * * * * * * * * * * * * * * * * * * * * *
kbget           proc    near                    ;kbd interrupt routine
                push    bx
                cli                             ;interrupt back off
                mov     bx,bufpt1               ;get pointer to head
                cmp     bx,bufpt2               ;test empty of buffer
                jnz     kbget2                  ;no,fetch a character
                cmp     kbflag,0
                jnz     kbget3
                sti                             ;allow an interrupt to occur
                pop     bx
                jmp     kbget                   ;loop until something in buff
kbget2:
                mov     al,[buffer+bx]          ;get ascii code
                inc     bx                      ;inc a buffer pointer
                cmp     bx,16                   ;at end of buffer ?
                jc      kbget3                  ;no,continue
                mov     bx,0                    ;reset to buf beginning
kbget3:
                mov     bufpt1,bx               ;store value in variable
                pop     bx
                ret                             ;return to main
kbget           endp
```

```
;————————————————————————————————
kbint       proc    far             ;keyboard interrupt routine
            push    bx
            push    ax

            in      al,60h          ;read in the character
            push    ax              ;save it
            in      al,61h          ;get the control port
            or      al,80h          ;set acknowledge bit for kbd
            out     61h,al
            and     al,7fh          ;reset acknowledge bit
            out     61h,al

            pop     ax              ;recover scan code
            test    al,80h          ;is press or release code?
            jnz     kbint2          ;is release code,return
            mov     bx,offset scantab
            xlat    scantab         ;ascii code to AL
            cmp     al,0
            jnz     kbint4
            mov     kbflag,80h
            jmp     kbint2
kbint4:
            mov     bx,bufpt2       ;buffer tail pointer
            mov     [buffer+bx],al  ;ASCII fill in buffer
            inc     bx
            cmp     bx,16           ;is end of buffer?
            jc      kbint3          ;no
            mov     bx,0            ;reset to buf beginning
kbint3:
            cmp     bx,bufpt1       ;is buffer full?
            jz      kbint2          ;yes, lose character
            mov     bufpt2,bx       ;save buf tail pointer
kbint2:     cli
            mov     al,20h          ;end of interrupt
            out     20h,al
            pop     ax
            pop     bx
            sti
            iret                    ;interrupt return
kbint       endp
;————————————————————————————————
dispchar    proc    near            ;(AL)=displaying char.
            push    bx
            mov     bx,0
            mov     ah,0eh          ;display function
            int     10h             ;call video routine
            pop     bx
            ret
dispchar    endp
;————————————————————————————————
code        ends                    ;end of code segment
            end     start
```

图 3.17 例 3.9 的程序清单

这个程序虽然比较长,但仔细分析程序,发现其中有一部分和例3.6键盘处理演示程序一样。其实,例3.6是一个很好的中断程序的例子,它检测按键时产生的中断,并把按键的扫描码转换为ASCII码存入缓冲区buffer,这个工作和ROM 09类型的键盘中断例程所完成的工作大致相同,只是在解释各个键时作了简化,即只解释了可显示字符,如英文小写字母、数字及一些符号;其他特殊键,如组合键、双态键、功能键都未作解释,只作为字符码0来处理。例3.6也以中断程序的标准形式来编写,在主程序部分保存原09类型的中断向量,设置自编例行程序的中断向量,设置中断屏蔽位,然后开中断。当键盘中断处理后,主程序中安排指令读取缓冲区中的字符并显示在屏幕上。在主程序的末尾,利用DOS功能恢复原09类型的中断向量,然后返回DOS。

在本例打字计时程序中,其键盘中断部分就采用现成的例3.6键盘处理程序,因此输入的字符也限于小写字母、数字及下挡符号,并且按动其他字符码为0的键会使程序结束,返回DOS。

打字的计时部分就是定时器中断处理部分。定时器的中断类型为08H,中断优先级比键盘高,只要在键盘中断处理程序KBINT中用一条STI指令允许高级中断,那么在整个键盘的输入过程中,定时器会以每秒18.2次的频率产生中断。根据计时的需要,在整个打字的过程中,必须把定时器中断的次数转换成时间,因此需要使用1CH中断扩充原定时器例行程序的功能。在自编的1CH中断例行程序中,定时器中断的次数记录在计数单元count中,当count计数值为18时,sec计数单元加1,当sec计数值达到60时,min计数单元加1。当然,输入一个句子无论如何也不会达到小时级,所以对hours计数值的调整判断,实际上只起到退出定时器中断的作用。

为了便于阅读,整个程序按两个中断源分为两部分。在数据段中,前半部分是键盘处理所需要的数据变量,包括字符转换表,字符缓冲区及其头尾指针,09中断向量的保存单元等。后半部分是定时器中断处理所需要的数据变量,其中有5个打字例句及其地址表saddr,计时单元count、sec、min、hours以及1CH中断向量的存储单元等。

在主程序main中,除对09类型的中断向量保存、设置、恢复外,还增加了对1CH类型的中断向量的保存、设置和恢复。在调用子程序dispchar显示键入字符时,还判断字符是否为回车符(0DH),如为回车符,说明一个句子输入完毕,此时调用显示时间的子程序disptime,显示出打字时间。然后顺序显示下一个例句,如果5个例句都显示完,则又从第一个例句开始显示,直至键入一个功能键退出程序,结束打字练习。

键盘中断处理程序kbint,对释放键不做处理,对按键进行转换,并将其ASCII码存入循环缓冲区buffer。

定时器中断处理程序clint的起始处将DS置为用户数据段data,这是因为产生08类型中断后,INT 08例行程序将DS置为ROM数据区,而在clint处理程序中,所引用的数据变量count、sec、min都在用户数据段中,所以必须将DS切换为data的段地址。在调整时间值时调用了一个子程序inct,使每18次中断计为1秒,每60秒计为1分,最后min,sec,count等计时单元中均保存着二进制表示的时间值。

子程序disptime分别将各计时单元的二进制数转换为十进制数,并以min:sec:msec的形式显示出来。其中msec是由count计数值转换成的:

$$\text{count} \times 1\text{秒}/18.2 = \text{count} \times 55\text{ms}$$

下面是打字练习程序的运行情况:

C：\type_ ex↙
  * PLEASE PRACTISE TYPING *
abcd efgh ijkl mnop qrst uvwx yz.
abcd efgh ijkl mnop qrst uvwx yz↙
001：002：605

christmas is a time of joy and love.
christmas is a time of joy and love.↙
000：054：880

store windows hold togs and gifts.
store windows hold togs and gifts.↙
000：044：580

people send christmas cards and gifts.
people send christmas cards and gifts.↙
000：042：550

santa wish all people peace on earth.
santa wish all people peace on earth.↙
000：038：660

— Esc 键退出
C：\

二、实验题

实验 3.8   中断练习程序

1. 题目:中断练习程序

2. 实验要求:存储器中有一个首地址为 BUFFER 的缓冲区,存放着一串 ASCII 码字符。要求编制实现以下功能的中断程序:在主程序运行期间,每 5 秒钟响铃一次;当键盘上的某个键被按下时,主程序和响铃都被挂起,显示器显示 BUFFER 缓冲区中的字符串,然后等待下一次按键引起的键盘中断;当键盘中断发生后,恢复主程序和响铃。这一过程可以重复任意次。

3. 提示

(1) 本实验需要定时器及键盘两个中断源,这两个中断源的关系是:在主程序运行期间(可用 LOOP 指令作空闲循环来模拟),CPU 既能响应定时器每秒 18.2 次的中断请求,也能响应键盘的中断请求。当第一次按键产生中断后,应禁止定时器的中断,并开始显示字符串。只有第二次按键产生中断时,才恢复定时器的中断并返回主程序。

(2) 主程序中应分别保存定时器及键盘的原中断向量,设置自编处理程序的中断向量,清除定时器和键盘的中断屏蔽位并开中断。在返回 DOS 之前,恢复定时器和键盘原来的中断向量。

(3) 按下键和放开键都能引起键盘中断,但在处理键盘中断时,对按键所产生的代码不必解释处理,只需根据读取的扫描码的最高位确定是按键中断还是释放键中断。如果是释放键引起的中断,则无须做任何工作,直接从中断处理程序中退出。如果是按下键引起的中断,则要区别是第一次按键还是第二次按键,以便作出不同的处理,为此可设置一个标志变量 flag。

(4)每次按下键产生的中断,使 flag 的最低位发生一次变化,设 flag 的初始值为 0,则第一次按键使其变为 1,第二次按键使其变为 0……,这样通过判断 flag 为 1 或 0 来区别两次按键,并分别转入不同的处理:

```
            in      al,60h          ;读键盘
            push    ax              ;保存扫描码
            in      al,61h
            or      al,80h
            out     61h,al          ;置键盘应答位
            in      al,61h
            and     al,7fh
            out     61h,al          ;复位键盘应答位
            pop     ax
            test    al,80h          ;是通码?
            jnz     inkret          ;不是,中断返回
            xor     flag,1          ;是通码,则触发标志位
            cmp     flag,1          ;第 1 次按键?
            je      process1        ;是,则挂起主程序和响铃
process2:   …                       ;是第二次按键,则恢复主程序及响铃
process1:   …                       ;挂起主程序和响铃

inkret:     mov     al,20h          ;中断结束命令
            mov     20h,al          ;送中断命令寄存器
            sti
            iret
```

(5)第一次按键的处理功能是屏蔽定时器中断,使之不再控制响铃,然后等待第二次键盘中断,为此必须清除第一次按键产生的中断级,否则将禁止第二次键盘中断。

```
process1:   in      al,21h
            or      al,01
            out     21h,al          ;屏蔽定时器中断
            call    display_char    ;显示字符串
            mov     al,20h          ;结束第 1 次键盘中断
            out     20h,al
            sti                     ;允许再次中断
again:      cmp     flag,1          ;等待第二次键盘中断
            je      again
```

第四章 文件管理

4.1 文件代号方式下的文件管理

一、示例

例 4.1 分页显示文件 ex_41

试编写一个程序以分页方式列出指定文件的内容,且可在显示过程中改变页的大小。根据上述要求,我们可以确定出该程序的用户界面:

(1) 在出现提示后输入要显示的文件的路径名;
(2) 显示一页后暂停,等待用户命令;
(3) 用户键入空格表示继续显示下一页;
(4) 用户键入 P 表示要改变以后每页的行数,这时程序打出提示行"Page Size:",等待用户键入一个整数;
(5) 每页的缺省行数为 24;
(6) 用户在 P 命令下键入的整数应在 1~24 之内,不在此范围者无效。

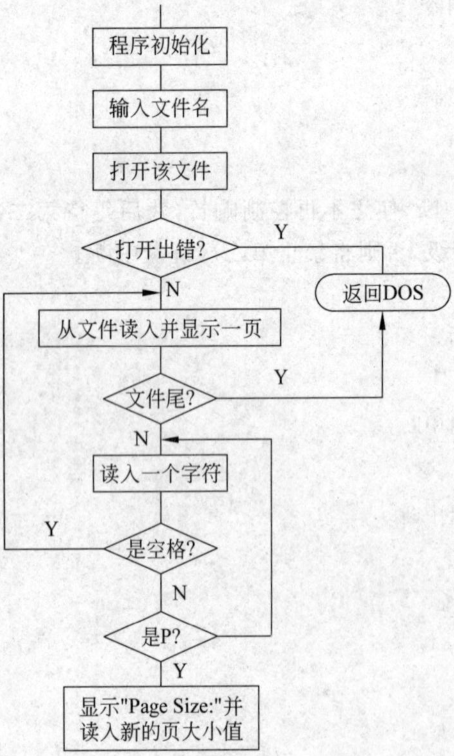

该程序至少需要以下几类变量(数据存储单元):

(1) 出错信息和输入提示信息变量;
(2) 文件名变量;
(3) 文件代号变量;
(4) 从文件读入信息时的临时存放区。

显然,该程序必须显示整个文件,因而是以显示至文件尾为退出条件的循环程序,每执行一次循环要等待用户的继续显示命令(空格),否则一直等待。由此我们很容易得出该程序的框图,如图 4.1 所示。

该程序编写时的难点在"从文件读入并显示一页。"因为每行字符数不等,每次读入的字符数又是固定的,就可能出现读入的字符不够一页甚至断在一行的中间,所以处理起来要小心。事实上,这个问题从另一角度描述一下也许更清楚一些:一个子程序 A 能够从指定文件中读入 200 个字符,放入以 buf 为首地址的内存缓冲区中,另一子程序 B 能够从 buf 中指定的地方显示一个字符,如何才

图 4.1 程序总框图

能将 A 和 B 恰当地协调起来,以显示出给定的行数呢? 图 4.2 的框图给出了一个解决办法。

图 4.2 的框图仅表达了显示一页的处理流程,其细节可参照给出的源程序(图 4.3)仔细阅读。该程序的运行方法与前面给出的用户界面是一致的,图 4.4 显示了一个简单的运行实例。

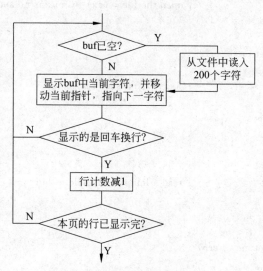

图 4.2 从文件读入并显示一页

```
    data segment
        Pgsize      dw      ?
        buf_size    db      80
        s_buf       db      ?
        buf         db      200  dup(?)
        cur         dw      ?
        handle      dw      ?
        mess_getname db     0dh,0ah,"    Please input filename: $ "
        mess_err1   db      0ah,0dh,"    Illegal filename ! $ "
        mess_err2   db      0ah,0dh,"    File not found ! $ "
        mess_err3   db      0ah,0dh,"    File read error ! $ "
        mess_psize  db      0ah,0dh,"    Page Size : $ "
        crlf        db      0ah,0dh," $ "
        mess_star   db      0ah,0dh," * * * * * * * * * * * * * * * * * * * "
                    db      0ah,0dh," $ "
    data    ends

    code segment
        assume ds:data, cs:code
        main proc far
    start:
        push    ds
        sub     ax,ax
        push    ax
```

```
        mov     ax,data
        mov     ds,ax

        mov     PgSize,12           ; Each page 12 lines.
        mov     cur,200             ; File data buffer is empty.
        call    getline             ; Get file name.
        call    openf               ; open the file, (ax)=0 means no such file
        or      ax,ax
        jnz     display
        mov     dx,offset mess_err2
        mov     ah,09h
        int     21h                 ;(ax)=0: no such file.

        jmp     file_end
display:
        mov     cx,PgSize
show_page:
        call    read_block          ; read a line from handle to buf
        or      ax,ax
        jnz     next2

        mov     dx,offset mess_err3
        mov     ah,09h
        int     21h                 ;(ax)=0: error in read.
        jmp     file_end
next2:
        call    show_block          ; display a line in buf,(bx) returned 0
                                    ; means that the file reach its end.
        or      bx,bx
        jz      file_end            ;(bx)=0: at the end of file.
        or      cx,cx
        jnz     show_page           ;(cx)<>0: not the last line of a page.
        mov     dx,offset mess_star
        mov     ah,09h
        int     21h                 ;At the end of a page, print a line of *.

; the current page has been on screen, and followed by a line of stars.

; the following part get the command from keyboard:
wait_space:
        mov     ah,1
        int     21h
        cmp     al," "
        jnz     psize
        jmp     display
psize:
        cmp     al,"p"
        jnz     wait_space
        call    change_psize
```

```
here:
        mov     ah,1
        int     21h
        cmp     al," "
        jnz     here                    ; stick here to wait for space.
        jmp     display

file_end:
        ret
main endp
;* * * * * * * * * * * * * * * * * * * * * * * * * * * *

;* * * * * * * * * * * * * * * * * * * * * * * * * * *
change_psize proc near
        push    ax
        push    bx
        push    cx
        push    dx
        mov     dx,offset mess_psize
        mov     ah,09h
        int     21h                     ;print the promt line

        mov     ah,01
        int     21h                     ;get the new num. of page size.
        cmp     al,0dh
        jz      illeg
        sub     al,"0"
        mov     cl,al
getp:
        mov     ah,1
        int     21h
        cmp     al,0dh
        jz      pgot
        sub     al,"0"
        mov     dl,al
        mov     al,cl
        mov     cl,dl                   ; exchange al and cl.

        mov     bl,10
        mul     bl
        add     cl,al
        jmp     getp
pgot:
        mov     dl,0ah
        mov     ah,2
        int     21h                     ; output 0ah to complete the RETURN.

        cmp     cx,0
        jle     illeg
        cmp     cx,24
        jg      illeg
        mov     PgSize,cx               ;be sure the new page size in (0..24) region.
illeg:
```

```
        mov     dl,0ah
        mov     ah,2
        int     21h             ;output 0ah to complete the RETURN.
        pop     dx
        pop     cx
        pop     bx
        pop     ax
        ret
change_psize endp
;* * * * * * * * * * * * * * * * * * * * * * * * * * * * * *

;* * * * * * * * * * * * * * * * * * * * * * * * * * * * * *
openf proc near
        push    bx
        push    cx
        push    dx
        mov     dx,offset buf
        mov     al,0
        mov     ah,3dh
        int 21h
        mov     handle,ax
        mov     ax,1
        jnc     ok
        mov     ax,0
ok:
        pop     dx
        pop     cx
        pop     bx
        ret
openf   endp
;* * * * * * * * * * * * * * * * * * * * * * * * * * * * * *

;* * * * * * * * * * * * * * * * * * * * * * * * * * * * * *
getline proc near
        push    ax
        push    bx
        push    cx
        push    dx
        mov     dx,offset mess_getname
        mov     ah,09h
        int     21h             ; promt user to input file name.

        mov     dx,offset buf_size
        mov     ah,0ah
        int     21h             ; function call of buffer input.

        mov     dx, offset crlf
        mov     ah,09h
        int     21h             ; return.

        mov     bl,s_buf
        mov     bh,0
        mov     [buf+bx],0      ; put 0 into the end of file name
```

```
        pop     dx
        pop     cx
        pop     bx
        pop     ax
        ret
getline endp
;* * * * * * * * * * * * * * * * * * * * * * * * * * * * *

;* * * * * * * * * * * * * * * * * * * * * * * * * * * * *
read_block proc near
        push    bx
        push    cx
        push    dx
        cmp     cur,200
        jnz     back

; if no more chars in buf can be displayed
; then read another 200 chars:

        mov     cx,200
        mov     bx,handle
        mov     dx, offset buf
        mov     ah, 3fh
        int     21h
        mov     cur,0
        mov     ax,1
        jnc     back
        mov     cur,200
        mov     ax,0
back:
        pop     dx
        pop     cx
        pop     bx
        ret
read_block endp
;* * * * * * * * * * * * * * * * * * * * * * * * * * * * *

;* * * * * * * * * * * * * * * * * * * * * * * * * * * * *
show_block proc near
        push    ax
        push    dx
        mov     bx,cur
loop1:
        cmp     bx,200
        jl      lp
        jmp     exit            ;if buf is empty then return.
lp:
        mov     dl,buf[bx]      ;else show the currnet char.
        cmp     dl,1ah          ; search the file end
        jz      exit_eof
        inc     bx
        inc     cur
```

```
            mov     ah,02
            int     21h
            cmp     dl,0ah
            jz      exit_ln          ;if the char shown is RETURN,
                                     ;then exit. A line has been on screen.
            jmp     loop1
exit_eof:
            mov     bx,0
exit_ln:
            dec     cx
exit:
            pop     dx
            pop     ax
            ret
show_block endp
;************************************
code ends
end start
```

图 4.3 例 4.1 的源程序

Please input filename: mac.lib

```
;********************************************
; The macros for label producing
;       ――― using stack to store labels
;********************************************

connect macro   x1,x2,x3,x4,x5
ifb     <x3>
        &x1&x2      &x4&x5
else
&x1&x2&x3&x4&x5


;********************************************
    endif
endm

labeling macro top
    connect ?_?,%s&top,:,
endm

branch macro top,con
    connect j,con,,?_?,%s&top
endm

st_asgn macro top,value

;********************************************
    connect s&top,=,value,,
endm
```

```
st_push  macro
    spointer=spointer+1
    counter=counter+1
    st_asgn %spointer,%counter
endm

st_pop   macro
    spointer=spointer-1
endm

* * * * * * * * * * * * * * * * * * * * * * * * * * * * * * * * *
p
    Page Size : 5

compares macro x1,op,x2
    ifidn  <x1>,<ax>
    cmp  x1,x2
    else
* * * * * * * * * * * * * * * * * * * * * * * * * * * * * * * * *
        ifidn  <x2>,<ax>
    cmp    x2,x1
        else
    push   ax
    mov    ax,x1
* * * * * * * * * * * * * * * * * * * * * * * * * * * * * * * * *
    cmp    ax,x2
    pop    ax
        endif
    endif
endm
* * * * * * * * * * * * * * * * * * * * * * * * * * * * * * * * *
p
    Page Size : 21

while   macro    x1,op,x2
local   next
    st_push
    st_push
    sptmp=spointer-1
    labeling   %sptmp            ; produce the loop head label.
    compares   x1,op,x2
    connect j,op,,next
    branch   %spointer,mp        ; jump out of the loop.
next:
endm

wend macro
    sptmp=spointer-1
    branch    %sptmp,mp
        ; jump to the beginning of the loop.
    labeling %spointer
        ; produce the loop exit label.
```

```
        st_pop
        st_pop
    endm

    * * * * * * * * * * * * * * * * * * * * * * * * * * * * * * * * *
    iff   macro x1,op,x2
        local nxt
        st_push
        sptmp=spointer
        st_push
        compares x1,op,x2
        connect j,op,,nxt ;to if clause
        branch %sptmp,mp ;branch to else clause
    endm

    elsee macro
        branch %spointer,mp ;to the end of whole if.
        labeling %sptmp
            ;produce the label of else clause.
    endm

    ifend macro
        labeling %spointer
        st_pop
        sy_pop
    endm

    * * * * * * * * * * * * * * * * * * * * * * * * * * * * * * * * *
    init_macs macro
        spointer=1
        counter=100
    endm
```

图 4.4 例 4.1 的运行实例

例 4.2 删除页 ex_42

为例 4.1 的程序增加删除命令，即在显示一页后，可接受 3 个命令：空格、P 和 D，其中 D 表示将刚才显示的一页从文件中删去，在接到空格命令之前，可以执行多次 P 命令和 D 命令。

要删除文件中一部分，简单的办法就是将需保留的部分写入另一临时文件，然后将该临时文件再写回原文件。

由于是在显示完某页后才知道是否删除该页，因此在显示该页中每一行的同时，应将该行保留下来，这样就可以在用户键入 D 或其他命令时决定是否将保存起来的该页写入临时文件。为保存一页的信息，必须设置一个大缓冲区 buf_tmp，它须有 24×80 个字节，这是一页 24 行，每行 80 个字符的标准设计的最大页空间。

页的保存很简单，只须在显示的同时将字符送入 buf_tmp 即可，当然，应记下该页字符总数，以便向临时文件写入。该程序的框图见图 4.5。

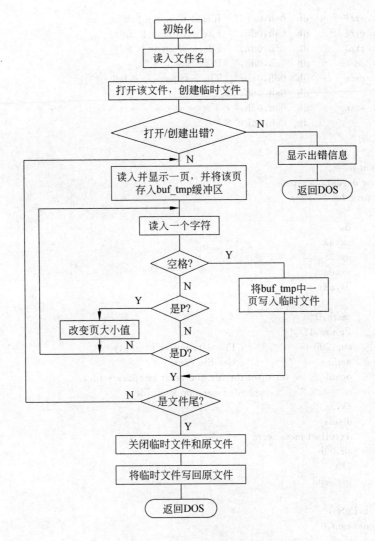

图 4.5 例 4.2 的程序框图

该例子的源程序及一个运行实例请见图 4.6 和图 4.7。

```
data    segment
    PgSize        dw      ?
    buf_size      db      80
    s_buf         db      ?
    buf           db      200     dup(?)
    names         db      20      dup(?)
    cur           dw      ?
    handle        dw      ?
    buf_tmp       db      24*80 dup(?)
    cur_tmp       dw      ?
    name_tmp      db      "t0m1p",0
    handle_tmp    dw      ?
    mark          db      ?
    mess_getname  db      0dh,0ah,"   Please input filename: $ "
```

```
        mess_err1      db    0ah,0dh," Illegal filename ! $"
        mess_err2      db    0ah,0dh," File not found ! $"
        mess_err3      db    0ah,0dh," File read error ! $"
        mess_psize     db    0ah,0dh," Page Size : $"
        mess_dele      db    0dh,0ah," The last page is deleted !"
        crlf           db    0ah,0dh,"$"
        mess_star      db    0ah,0dh,"* * * * * * * * * * * * * * * * * * * * * * *"
                       db    0ah,0dh,"$"
data ends

code segment
        assume ds:data, cs:code
        main proc far
start:
        push    ds
        sub     ax,ax
        push    ax
        mov     ax,data
        mov     ds,ax

        mov     mark,0
        mov     PgSize,12
        mov     cur,200              ; File data buffer is empty
        call    getline              ; Get file name
        call    openf                ; open the file and creat temporary file,
                                     ; (ax)=0 means no such file
        or      ax,ax
        jnz     display
        mov     dx,offset mess_err2
        mov     ah,09h
        int     21h
        jmp     file_end
display:
        mov     cx,PgSize
        mov     cur_tmp,0
show_page:
        call    read_block           ; read a line from handle to buf
        or      ax,ax
        jnz     next2
        mov     dx,offset mess_err3
        mov     ah,09h
        int     21h                  ; error in read.
        jmp     file_end
next2:
        call    show_and_reserve     ; display a line in buf ,
                                     ; and put the line in buf_tmp.
                                     ; (bx) returned = 0
                                     ; means that the file reach its end.
        or      bx,bx
        jz      file_end             ;(bx)=0 : at the end of file.

        or      cx,cx
        jnz     show_page
```

; (cx)=0 ; end of a page, print a line of stars.

```
        mov     dx,offset mess_star
        mov     ah,09h
        int     21h
```

; the current page has been on screen,
; and followed by a line of stars.

```
wait_space:
        mov     ah,1
        int     21h
        cmp     al," "
        jnz     psize
        call    write_buf_tmp
; command=space , then reserve the page in temp file.
        jmp     display
psize:
        cmp     al,"p"
        jnz     delete
        call    write_buf_tmp
```

; the last page is reserved.

```
        call    change_psize
        jmp     stick
delete:
        cmp     al,"d"
        jnz     wait_space
```

; command is DELETE ,the last page
; not reserved in temp file.

```
        mov     mark,1
        mov     dx,offset mess_dele
        mov     ah,09h
        int     21h
stick:
        mov     ah,1
        int     21h
        cmp     al," "
        jnz     stick
        jmp     display
file_end:
        call    write_buf_tmp       ; reserve the last page in
                                    ; temp file.
        cmp     mark,0
        jz      ok
        call    write_tmp_back      ; write the temp file back
                                    ; to user's file.
ok:
        ret
main endp
```
;*************************

```
;************************
change_psize proc near
        push    ax
        push    bx
        push    cx
        push    dx
        mov     dx,offset mess_psize
        mov     ah,09h
        int     21h
        mov     ah,01
        int     21h

        cmp     al,0dh
        jz      illeg
        sub     al,"0"
        mov     cl,al
getp:
        mov     ah,1
        int     21h
        cmp     al,0dh
        jz      pgot
        sub     al,"0"
        mov     dl,al
        mov     al,cl
        mov     cl,dl           ; exchange al and cl.

        mov     bl,10
        mul     bl
        add     cl,al
        jmp     getp
pgot:
        mov     dl,0ah
        mov     ah,2
        int     21h             ; output 0ah to complete the RETURN.
        cmp     cx,0
        jle     illeg
        cmp     cx,24
        jg      illeg
        mov     PgSize,cx
illeg:
        mov     dl,0ah
        mov     ah,09h
        int     21h             ; output 0ah to complete the RETURN.
        pop     dx
        pop     cx
        pop     bx
        pop     ax
        ret
change_psize endp
;************************

;************************
```

```
openf proc near
        push    bx
        push    cx
        push    dx
        mov     dx,offset names
        mov     al,2
        mov     ah,3dh
        int     21h
        mov     handle,ax
        mov     ax,0
        jc      quit
        mov     dx,offset name_tmp
        mov     cx,0
        mov     ah,3ch

        int     21h
        mov     handle_tmp,ax
        jc      quit
        mov     ax,1
quit:
        pop     dx
        pop     cx
        pop     bx
        ret
openf endp
;* * * * * * * * * * * * * * * * * * * * * * * * * * *

;* * * * * * * * * * * * * * * * * * * * * * * * * * *
getline proc near
        push    ax
        push    bx
        push    cx
        push    dx
        mov     dx,offset mess_getname
        mov     ah,09h
        int     21h
        mov     dx,offset buf_size
        mov     ah,0ah
        int     21h
        mov     dx, offset crlf
        mov     ah,09h
        int     21h
        mov     bl,s_buf
        mov     bh,0
        mov     names[bx],0     ; insert 0 to form the asciiz string.
name_move:
        dec     bx
        mov     al,buf[bx]
        mov     names[bx],al    ; move the line got into name string
        jnz     name_move
        pop     dx
        pop     cx
        pop     bx
```

```
            pop     ax
            ret
getline endp
;* * * * * * * * * * * * * * * * * * * * * * * * * *

;* * * * * * * * * * * * * * * * * * * * * * * * * *
read_block proc near
            push    bx
            push    cx
            push    dx
            mov     ax,1
            cmp     cur,200
            jnz     back
            mov     cx,200
            mov     bx,handle
            mov     dx, offset buf
            mov     ah, 3fh
            int     21h
            mov     cur,0
            mov     ax,1
            jnc     back
            mov     cur,200
            mov     ax,0
back:
            pop     dx
            pop     cx
            pop     bx
            ret
read_block endp
;* * * * * * * * * * * * * * * * * * * * * * * * * *

;* * * * * * * * * * * * * * * * * * * * * * * * * *
show_and_reserve proc near
            push    ax
            push    dx
            mov     bx,cur
            mov     bp,cur_tmp
loop1:
            cmp     bx,200
            jl      lp
            jmp     exit
lp:
            mov     dl,buf[bx]
            mov     ds:buf_tmp[bp],dl   ; (dl) need shown, reserve
                                        ; it in buf_tmp.
            inc     bx
            inc     cur
            inc     bp
            inc     cur_tmp

            cmp     dl,1ah      ; search the file end
            jz      exit_eof
            mov     ah,02
```

```
        int     21h             ; show the (dl).
        cmp     dl,0ah
        jz      exit_ln         ; if meets RETURN, exit.
        jmp     loop1           ; else show another char.
exit_eof:
        mov     bx,0
exit_ln:
        dec     cx
exit:
        pop     dx
        pop     ax
        ret
show_and_reserve endp
;* * * * * * * * * * * * * * * * * * * * * * * * * *

;* * * * * * * * * * * * * * * * * * * * * * * * * *
write_buf_tmp proc near
        push    ax
        push    bx
        push    cx
        push    dx
        mov     dx,offset buf_tmp
        mov     cx,cur_tmp
        mov     bx,handle_tmp
        mov     ah,40h
        int     21h
        pop     dx
        pop     cx
        pop     bx
        pop     ax
        ret
write_buf_tmp endp
;* * * * * * * * * * * * * * * * * * * * * * * * * *

;* * * * * * * * * * * * * * * * * * * * * * * * * *
write_tmp_back proc near
        push    ax
        push    bx
        push    cx
        push    dx
        mov     bx,handle_tmp
        mov     ah,3eh
        int     21h             ; close the temporary file.
        mov     bx,handle
        mov     ah,3eh
        int     21h             ; close the file giving.
        mov     dx,offset name_tmp
        mov     al,0
        mov     ah,3dh
        int     21h             ; open the temporary file for reading.
        mov     handle_tmp,ax
        mov     dx,offset names
```

```
            mov     al,1
            mov     ah,3dh
            int     21h                ; reopen the file giving for writing.
            mov     handle,ax
            mov     si,1
wrt_back:
            mov     bx,handle_tmp
            mov     ah,3fh
            mov     cx,200
            mov     dx,offset buf
            int     21h                ; read a page bytes from temporary file to buf.
            jc      wrt_end
            mov     si,ax
            mov     bx,handle
            mov     ah,40h
            mov     cx,200
            mov     dx,offset buf
            int     21h                ; write a page bytes from buf to the file giving.
            or      si,si
            jnz     wrt_back
            mov     ah,3eh
            mov     bx,handle
            int     21h                ; close the file giving.
wrt_end:
            pop     dx
            pop     cx
            pop     bx
            pop     ax
            ret
write_tmp_back endp
;* * * * * * * * * * * * * * * * * * * * * * * * * *
code ends
end start
```

图 4.6 例 4.2 的源程序

```
    Please input filename: m


;* * * * * * * * * * * * * * * * * * * * * * * * * *
; The macros for label producing
;    ——— using stack to store labels
;* * * * * * * * * * * * * * * * * * * * * * * * * *
connect macro   x1,x2,x3,x4,x5
ifb <x3>
            &x1&x2       &x4&x5
else
&x1&x2&x3&x4&x5
* * * * * * * * * * * * * * * * * * * * * * * * * *
    endif
endm
```

```
labeling macro top
        connect ?_?,%s&top,:,
endm

branch macro top,con
        connect j,con,,?_?,%s&top
endm
st_asgn macro top,value
   * * * * * * * * * * * * * * * * * * * * * * * * * * * * *
p
    Page Size : 6
        connect    s&top,=,value,,
endm

st_push macro
        spointer=spointer+1
        counter=counter+1
   * * * * * * * * * * * * * * * * * * * * * * * * * * * * *
        st_asgn    %spointer,%counter
endm
st_pop   macro
        spointer=spointer-1
endm
   * * * * * * * * * * * * * * * * * * * * * * * * * * * * *
d
        The last page is deleted !

compares macro x1,op,x2
    ifidn <x1>,<ax>
    cmp x1,x2
    else
        ifidn <x2>,<ax>
   * * * * * * * * * * * * * * * * * * * * * * * * * * * * *
    cmp x2,x1
        else
    push ax
    mov ax,x1
    cmp ax,x2
    pop ax
   * * * * * * * * * * * * * * * * * * * * * * * * * * * * *
d
    The last page is deleted !
        endif
    endif
endm
while macro x1,op,x2
local next
   * * * * * * * * * * * * * * * * * * * * * * * * * * * * *
    st_push
    st_push
    sptmp=spointer-1
```

```
        labeling %sptmp         ;produce the loop head label.
        compares x1,op,x2
        connect j,op,,next
     * * * * * * * * * * * * * * * * * * * * * * * * * * * * * *
          branch    %spointer,mp    ;jump out of the loop.
next:
endm

wend macro
    sptmp=spointer-1
     * * * * * * * * * * * * * * * * * * * * * * * * * * * * * *
p
    Page Size : 12
        branch %sptmp,mp
            ;jump to the beginning of the loop.
        labeling %spointer
            ;produce the loop exit label.
        st_pop
        st_pop
endm

iff macro x1,op,x2
    local nxt
    st_push
    sptmp=spointer
    st_push
    compares x1,op,x2
     * * * * * * * * * * * * * * * * * * * * * * * * * * * * * *
        connect j,op,,nxt ;to if clause
     branch %sptmp,mp ;branch to else clause
endm

elsee macro
    branch %spointer,mp ;to the end of whole if.
        labeling %sptmp
         ;produce the label of else clause.
endm

ifend macro
    labeling %spointer
    st_pop
     * * * * * * * * * * * * * * * * * * * * * * * * * * * * * *
         sy_pop
endm

init_macs macro
    spointer=1
    counter=100
endm
```

图 4.7 例 4.2 的运行实例

二、实验题

实验 4.1　页拷贝

1. 题目:页拷贝
2. 要求:

为例 4.2 的程序增加一个页拷贝命令 C,具体要求如下:

(1) 在显示一页后暂停显示,可接受空格、P、D 和 C 命令。

(2) 在接到空格命令前,不会继续显示。

(3) 接到用户键入的 C 命令后,显示"Copy To:"并等待用户输入目的文件名,该文件名以回车换行结束。

(4) 接到目的文件名后,打开或创建(如果找不到该文件)该目的文件。然后将刚才显示的那一页拷至目的文件的尾部,与其原有内容相接。最后关闭该目的文件。

(5) 如果给出的文件名是空,则直接返回,不做任何动作。如果文件名非法,则在创建时给出错误信息,然后返回。

(6) 在显示一个文件的过程中,允许同一页拷至不同文件,也允许其中几页拷入同一文件。

(7) 同一页可以先用 D 命令删除,再用 C 命令拷入其他文件。

3. 提示:

该实验中关键问题是如何将 buf_tmp 中的一页信息准确地接在目的文件的尾部。简单的办法就是顺序读入目的文件的内容,直到文件尾,这时可以算出该文件的准确大小(包含的字节数)。设每次读入 m 个字节,共读入 n 次,在第 n 次读入的内容中包含文件结束标志 1AH,第一个 1AH 为第 n 次读入内容中第 k 个字节(从 1 算起),则该文件共包含$(n-1) \times m+k-1$ 个字节。在确定了文件大小后,用移动文件指针的办法,将文件指针移至第一个 1AH 处(从文件头移时,要求先关闭,重新打开,以使文件指针指向文件头),再将 buf-tmp 中的页写入该目的文件,最后关闭。

这种文件的续接在编程和调试时会很容易出问题,而且很难从程序中看出来。正确使用 Debug 对调试这个程序很有帮助。

在调试程序的过程中,你往往只用 type 或用 WS 命令查看被拷贝的文件,一旦发现应接在后面的页没有出现,就无法得到更详尽的信息。出现这种情况时,可以起动 debug 程序,在 debug 中用 n 命令将被拷贝的文件装入,然后用 d 命令依次显示该文件内容,直至查到 1AH 为至。这时你可以清楚地观察到应拷入的页与原目的文件尾的距离。有时你可能会发现那一页距 1AH 仅差一个字符,但在 type 时一遇到 1AH 就不会再向下显示而认为文件结束,这时你一定知道应该如何去修改你的程序了。

4.2　文件控制块方式下的文件管理

一、示例

例 4.3　个人档案文件管理 ex_43

设有一个人档案文件 DOCU,它包含每个人的姓名、年龄、性别、身高和体重这五项内容。其中姓名占 10 个字节,年龄 2 个字节,性别 1 个字节,身高 3 个字节,体重 2 个字节。试编写一 DOCU 的管理程序,它接受如下三个命令:

(1) L 命令:列出 DOCU 中所有人的情况,每人一行。
(2) I 命令:在 DOCU 文件最后插入一个人的情况(我们称为一个记录)。
(3) Q 命令:退出 DOCU 的管理程序。

管理程序用"DOCU>>"作为可接受用户输入上述命令的提示符。

根据上述要求,我们不难确定该程序的粗略框图。

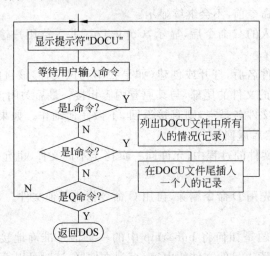

图 4.8 档案管理程序框图

从图 4.8 的框图可以看出,这实际是一种监控程序结构,只管辨别和接收命令,命令的详细处理没有给出,可以作为子程序实现。

L 命令的处理可由一个子程序完成,不妨叫 list-all,I 命令同样可以对应到一个子程序 insert,图 4.9 和图 4.10 描述了它们的基本流程。

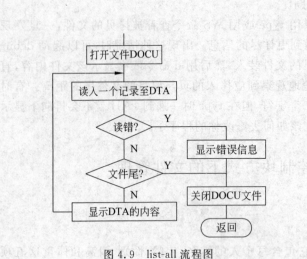
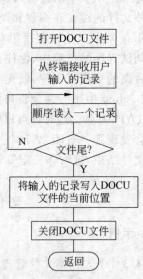

图 4.9 list-all 流程图　　　　图 4.10 insert 子程序框图

在具体编写程序之前,还有一些不可忽略的细节应考虑清楚,其中最要紧的是时刻牢记段前缀与数据段的差别,正确使用 DS 寄存器。为清楚起见,本程序在实现时将段前缀的说明部分单独作为一段,并将 DS 固定指向该段;其它数据单元(提示信息等)另设一段,以 ES 固定指向该段,在引用该段变量时加上 ES 段前缀即可。上述措施不是唯一的,仅作参考。

另外,文件名 DOCU 如何放入段前缀的 FCB 中也是一个问题。可有三种办法供选择:(1)在程序一开始用几条指令将 FCB 的文件各部分初始化为"DOCU";(2)从键盘读入"DOCU"再写入 FCB;(3)利用命令行参数直接初始化 FCB 的文件名。最简单的办法是第(3)种办法,只须在程序名后跟一个 DOCU 就行了,即在 DOS 状态下如下执行 DOCU 管理程序:

C>ex_43   DOCU↵

这也是我们常见的 DOS 命令处理参数的方法。

那么是不是只有文件控制块方式才能使用命令行参数法呢?不是的,系统在分析 DOS 的命令行后,将参数直接放入 FCB 的 5ch 处(第二个参数放入 6ch 处),因而可以从该处取出参数,进行各种处理。

最后需要说明的是,在编写 list_all 子程序时,增加了显示的分页功能,并且将一个记录的 5 项内容分开显示,以便于从屏幕阅读。同样,为方便用户,在 I 命令要求输入一个记录时,设计成 5 项分别输入并给出提示。总之,应尽量使你的程序有一个友好的用户界面。图 4.11 和图 4.12 给出了源程序和运行实例。

```
data segment
    org 6ah
    rec_size dw ?
    org 07ch
    recno db ?
    randlow dw ?
    randhi dw ?
    org 80h
    dta db 80h dup(?)
data ends

PSize    equ    8
RSize    equ    18
fcb      equ    5ch

vars segment
    names db 10 dup(?)
    age db ?,?
    sex db ?
    height db ?,?,?
    weight db ?,?
    pcounter db ?
    mess_n db 0ah,0dh,"        name        : $ "
    mess_a db 0ah,0dh,"        age         : $ "
```

```
        mess_s db 0ah,0dh,"      sex      : $ "
        mess_h db 0ah,0dh,"      height   : $ "
        mess_w db 0ah,0dh,"      weight   : $ "
        err_fopen  db 0ah,0dh,"    File open error !  $ "
        err_fcreat db 0ah,0dh,"    File creat error !  $ "
        err_fread1 db 0ah,0dh,"    File read error in sequence manner !  $ "
        err_fread2 db 0ah,0dh,"    File read error in random manner !  $ "
        err_fclose db 0ah,0dh,"    File close error !  $ "
        prompt db 0ah,0dh,"DOCU>> $ "
        continue db 0ah,0dh,"          Press space to continue... $ "
vars ends

show macro addrs
;   display a line started at addrs.

        push    ds
        mov     ax,es
        mov     ds,ax
        lea     dx,addrs
        mov     ah,09h
        int     21h
        pop     ds
endm

show_item macro addrs,count1,count2
local iloop,spacelp
        mov     bx,0

; display (count1) chars begin at addrs, and
; followed by (count2) spaces.

iloop:
        mov     dl,es:&addrs&[bx]
        mov     ah,2
        int     21h
        inc     bx
        cmp     bx,count1
        jl      iloop
        mov     cx,count2
spacelp:
        mov     dl," "
        mov     ah,2
        int     21h
        loop    spacelp
endm

getin macro addrs,count
    local lp,zeroit,input_end,exit
        push    ds
        mov     ax,es
        mov     ds,ax
```

```
        mov     bx,0

; get (count) chars from keyboard, and put them
; to addrs.

zeroit:
        mov     &.addrs&.[bx],0
        inc     bx
        cmp     bx,count
        jl      zeroit              ;let (count) num. of byte
                                    ; starting at addrs be zero.

        mov     bx,0
lp:     mov     ah,1
        int     21h
        cmp     al,0dh
        jz      input_end
        cmp     al,0ah
        jz      input_end
        mov     &.addrs&.[bx],al
        inc     bx
        cmp     bx,count
        jl      lp
input_end:
        cmp     al,0dh
        jz      exit
        cmp     al,0ah
        jz      exit

        mov     ah,1
        int     21h
        jmp     input_end           ; if not all char typed in have
                                    ; been received, then get again.
exit:
        pop     ds
endm

code segment
assume cs:code,ds:data,es:vars
main proc far
start:
        push    ds
        sub     ax,ax
        push    ax
        mov     ax,vars
        mov     es,ax

main_loop:
        show    prompt

; receive commands (l,i,q) and call the
```

```
; responding subroutine.

cmd_loop:
    mov     ah,1
    int     21h
    cmp     al,"l"
    jnz     n1
    call    list_all
    jmp     main_loop
n1:
    cmp     al,"i"
    jnz     n2
    call    insert
    jmp     main_loop
n2:
    cmp     al,"q"
    jnz     cmd_loop
    ret
main endp

;******************************
; list_all procedure lists the document        *
; file in a page manner                        *
;******************************
list_all proc near
    push    ax
    push    bx
    push    cx
    push    dx
    call    openf                   ; Open the DOCU file
    or      al,al
    jz      go_on
    jmp     ext
go_on:
    mov     recno,0
    mov     rec_size,RSize
list_loop:
    mov     dx,fcb
    mov     ah,14h
    int     21h                     ; read a record in sequence manner.
    cmp     al,01
    jne     n3
    jmp     file_end                ; (al)=1: meets the end of file.
n3:
    cmp     al,0
    je      n4
    show    err_fread1              ; read errror
    mov     dl,"0"
    add     dl,al
    mov     ah,2
    int     21h                     ; show error no.
```

```
        jmp     file_end
n4:
                                ; reading correct.
        mov     bx,0
moving:
                                ; moving the record to NAMES in
                                ; DS segment.

        mov     al,dta[bx]
        mov     es:names[bx],al
        inc     bx
        cmp     bx,RSize
        jl      moving

        mov     dl,0dh
        mov     ah,2
        int     21h
        mov     dl,0ah
        mov     ah,2
        int     21h             ; show RETURN.

        call    show_rec        ; show the record.

        mov     dl,0dh
        mov     ah,2
        int     21h
        mov     dl,0ah
        mov     ah,2
        int     21h             ; show RETURN.

        inc     es:pcounter
        cmp     es:pcounter,PSize
        jl      list_loop       ; show record in page manner.

        mov     es:pcounter,0
        show    continue        ; pause at the end of a page,
                                ; and print the promt message
                                ; to wait for continue.
pg_cmd:
        mov     ah,1
        int     21h
        cmp     al," "
        jne     pg_cmd          ; wait for space to continue.

        jmp     list_loop
file_end:
        call    closef
ext:
        pop     dx
        pop     cx
        pop     bx
```

```
        pop     ax
        ret
list_all endp

show_rec proc near
        push    ax
        push    bx
        push    cx
        push    dx
        show_item names,10,2
        show_item age,2,4
        show_item sex,1,3
        show_item height,3,2
        show_item weight,2,2
        pop     dx
        pop     cx
        pop     bx
        pop     ax
        ret
show_rec endp
;******************************************
; The insert procedure get a record from        *
; terminal, and insert it into DOCU file.       *
;******************************************
insert proc near
        push    ax
        push    bx
        push    cx
        push    dx
        call    openf
        or      al,al
        jz      getrc
        jmp     back
getrc:
        call    get_record        ; get a record form keybaord.
        mov     recno,0
        mov     rec_size,RSize
find_end:
        mov     ah,14h
        mov     dx,fcb
        int     21h               ; read a record from DOCU.
        cmp     al,01
        je      write_record      ; when read to the end,
                                  ; write the record to the
                                  ; end of DOCU.
        cmp     al,0
        je      find_end
write_record:
        mov     bx,0
to_dta:
        mov     al,es:names[bx]
```

```
        mov     dta[bx],al
        inc     bx
        cmp     bx,RSize
        jl      to_dta
        mov     ah,15h
        mov     dx,fcb
        int     21h                         ; write arecord to DOCU.
        call    closef
back:
        pop     dx
        pop     cx
        pop     bx
        pop     ax
        ret
insert endp

get_record proc near
        push    ax
        push    bx
        push    cx
        push    dx
        show    mess_n
        getin   names,10
        show    mess_a
        getin   age,2
        show    mess_s
        getin   sex,1
        show    mess_h
        getin   height,3
        show    mess_w
        getin   weight,2
        pop     dx
        pop     cx
        pop     bx
        pop     ax
        ret
get_record endp

openf proc near

        push    bx
        push    cx
        push    dx
        mov     ah,0fh
        mov     dx,fcb
        int     21h
        or      al,al
        jz      found
        show    err_fopen
        mov     ah,16h
        mov     dx,fcb
```

```
        int     21h
        or      al,al
        jz      found
        show    err_fcreat
found:
        pop     dx
        pop     cx
        pop     bx
        ret
openf endp

closef proc near
        push    ax
        push    dx
        mov     ah,10h
        mov     dx,fcb
        int     21h
        pop     dx
        pop     ax
        ret
closef endp

code ends
end start
```

图 4.11　例 4.3 的源程序

```
DOCU>>l
    abc         12      f       123     56

DOCU>>i
    name    : zhang
    age     : 25
    sex     : m
    height  : 167
    weight  : 65
DOCU>>l
    abc         12      f       123     56

    zhang       25      m       167     65

DOCU>>i
    name    : wang
    age     : 56
    sex     : f
    height  : 187
    weight  : 45
DOCU>>l
    abc         12      f       123     56

    zhang       25      m       167     65

    wang        56      f       187     45
```

DOCU>> q

图 4.12　例 4.3 的运行实例

二、实验题

实验 4.2　个人档案管理系统

1. 题目：个人档案管理系统
2. 要求：

在例 4.3 的基础上增加两条命令：

(1) S 命令：按指定项的值查询。即用户指定 5 项中一项，并给出要查的值，由程序负责搜索 DOCU 文件中具有与给定值相同的项的记录，并显示这些记录。S 命令的用户接口如下所示：

```
DOCU>>S↙
        1. name
        2. age
        3. sex
        4. height
        5. weight
        key item:2↙
        key value:21↙
        ……
```

其中有下画线的为用户输入部分，其他由程序显示，输入的 2 表示第 2 项，21 表示要查的第 2 项的值是 21，最后应列出查询结果，即所有年龄为 21 的记录。

(2) D 命令：删去指定记录。待删记录的确定与 S 命令中要显示的记录的确定一样，也由指定项的值来实现，其用户接口也一样。(1) 中的接口例子在 D 命令中则表示删去所有年龄为 21 的记录。

3. 提示

在 S 和 D 命令中，最关键的在于用给定值去与读入记录中相应字段比较。由于每项长度不同，如何根据给定的项号(key item)准确地确定该项在一个记录中的位置，是一个比较棘手的问题。

假设一个记录中各项顺序依次是：name、age、sex、height 和 weight，如图 4.13 所示，则可预先确定出每项的起始位置(相对于记录头部)，我们可以根据这来设计一个项号与该项位置的对应关系表，如图 4.14 所示，该表又可表示为如下的数据段说明：

```
position    dw    0     ;name 的位置
            dw    10    ;age 的位置
            dw    12    ;sex 的位置
            dw    13    ;height 的位置
            dw    16    ;weight 的位置
```

如果从 DOCU 文件读入的记录起始地址为 dta，项号为 key，则该项的第一个字符的偏移地址是：[position+(key-1)×2]+dta，其中方括号表示取括号中表达式指定单元的内容，position 是偏移地址。

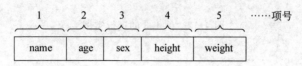

图 4.13 记录中项的次序安排

项号	初始位置
1	0
2	10
3	12
4	13
5	16

图 4.14 项号与起始位置的关系

4．提高要求

如果想使该档案系统更完善、功能更强，可以在 S 和 D 命令上作一些修改，使之可以查询或删除具有某一范围值的记录，如删除所有 19≤年龄＜23 的记录。若想更接近实用，则应提供多项综合查询和删除，即由多项的条件来确定相应记录，但难度相当大。

第五章　　高级汇编语言技术与连接技术

5.1　高级汇编语言技术

一、示例

例 5.1　用宏和高级汇编技术实现 if 和 while 语句功能 ex_51

试用宏和高级汇编语言技术实现类似高级语言中的循环控制和条件分支语句。为简单起见,循环语句 while 和条件分支语句 iff 的功能规定如下:

(1) while 后的循环条件为"x_1,op,x_2"形式,其中 x_1 和 x_2 为操作数,op 为关系比较符,用 g($>$)、l($<$)、e($=$)、ge($\geqslant$)、le($\leqslant$)表示。另外,x_1 和 x_2 必须是字。

(2) 循环结束用 wend 表示。

(3) iff 语句后条件的表达同 while。

(4) elsee 语句(相当于 else 语句)是可选项,即 iff 后可以不跟 elsee。

(5) 条件分支 iff 的结束用 ifend 表示。

(6) while 和 iff 允许相互嵌套,也允许自我嵌套。

从上述 6 个功能可以看出,这里的 while 和 iff 已相当接近高级语言了,由于允许嵌套,实现起来的难度确实不小。下面,我们从图 5.1 研究起。

```
        ...
    while  x₁,op,x₂
        ...
    wend
        ...
```

图 5.1　while 语句基本结构图

从图 5.1 可以看出,while 语句就是在 x_1,op,x_2 成立时循环执行它和 wend 之间的语句,我们据此可以将其转化成等价的汇编语言指令,如图 5.2 所示。

```
              ...
         ┌ label_0:
         │     cmp    x₁,x₂
  while ─┤     jop    label_1; 应为 jl 或 jg 等
         │     jmp    label_2
         └ label_1:

   wend ─┬     jmp    label_0
         └ label_2:
```

图 5.2　while 和 wend 的展开形式

如果我们有办法设计出两个正确的宏指令 while 和 wend,使得对它们的宏调用恰能完成由图 5.1 到图 5.2 的转变,那么这样的 while 和 wend 正符合我们的要求。如何做,这需要解决标号和比较、转移指令的自动生成问题,我们分别予以考虑。

首先,标号的自动生成问题必须解决。对自动生成的标号有一个最基本的要求,就是不能重复。为此,我们需设立一个标号计数器,通过利用宏具有的连接功能产生不同的标号,设标号计数器叫 spointer,产生标号的宏叫 labeling,形式如下:

```
        Spointer=0
        labeling  macro  top
            label_&top&:
                endm
```

对宏 labeling 作调用 labeling %spointer 就会产生一个字符串"label_0:",这正是一个合法的标号,下次只要在调用 labeling %spointer 之前作一次 spointer=spointer+1,即可得到一个新的标号"label_1:"。当然,为保险起见,可以产生一些更特殊的标号形式,如"?_?00:"。

另外一个问题是如何将 while 和 wend 转化为汇编指令。我们对照图 5.2 来看 while 这个宏应如何来写。首先,我们设计一个专门用来拼接的宏 connect,以便产生需要的汇编指令:

```
        connect    macro    x₁,x₂,x₃,x₄,x₅
            ifb    〈x3〉
                &x₁&x₂  &x₄&x₅
            else
                &x₁&x₂&x₃&x₄&x₅
            endif
                endm
```

该宏有两个用途:(i) x_3 为空时产生汇编指令,如 connect j,mp, label_,1 产生 jmp label_1;(ii) x_3 不为空时可以产生标号,如 connect A,B,C,1,:产生 ABC1:。

有了 connect 宏的支持,我们可以根据图 5.2 写出宏 while 的基本形式:

```
        while  macro   x₁,op,x₂
        labeling  %spointer      ;产生 label_0:
        cmp   x₁,x₂
        spointer=spointer+1
        connect  j,op,,label_,%spointer    ;产生 jop label_1
        spointer=spointer+1
        connect j,mp,,label_,%spointer     ;产生 jmp label_2
        spointer=spointer-1
        labeling %spointer                 ;产生 label_1:
            endm
```

类此,我们可以得到 wend 的宏的基本实现方法:

```
        wend macro
            spointer=spointer-1
            connect j,mp,,label_,%spointer    ;产生 jmp label_0
            spointer=spointer+2
            labeling   %spointer              ;产生 label_2:标号
            endm
```

如果不要求嵌套，只允许 while 语句的并列，那么上述 while 和 wend 已经达到目的了。但如果要求嵌套，那么 while 和 wend 之间将不能维持正确的标号和转移，因为从宏 wend 的实现来看，它产生的 jmp label_0 语句依赖于调用 wend 前的 spointer 值的正确性，即它应是与之相应的 while 调用后的 spointer 值，而在一对 while 和 wend 中间嵌入另一对 while 和 wend，恰好破坏了 spointer 的正确性。

问题的关键在于 while 和 wend 在 spointer 的值上必须一致，也就是说不管在该对 while 和 wend 中间嵌套了多少层 while 和 wend，应保证调用这个最外层 wend 时，spointer 的值应和没有这些嵌套时一样。这使我们不由联想起堆栈的性质，确实，我们正好需要一个堆栈来保存 spointer。

遗憾的是，我们不能简单地将 spointer 存入堆栈段，因为在汇编时无法取出堆栈段中的内容。不过，我们可以利用特殊的手段来达到目的：

```
        spointer=0
        counter=0
        st_asign macro top,value
            connect,s&top,=,value,,
                endm
        st_push macro
            spointer=spointer+1
            counter=counter+1
            st_asign %spointer,%counter
                endm
        st_pop macro
            spointer=spointer-1
                endm
```

上述宏定义中，st_asign 用来产生形如 s1=1、s2=2、s3=100、s4=101 之类的语句，这里的 s1、s2、s3 用来组成一个堆栈，用来存放标号的数字部分，该栈的进、出由 st_push 和 st_pop 控制，如图 5.3 所示。

利用标号栈的功能，可以在 while 中产生或引用一个新的标号时将其压入标号栈，当 wend 调用时再从标号栈中弹出相应标号。注意，这里真正的标号值是由 counter 确定的，spointer 是栈顶指针。

利用标号栈的功能对原 while 和 wend 稍加修改和细化，即可得到允许任意重嵌套和并列的 while 循环语句，同样可以很容易地得到 iff 语句的宏定义。具体的宏定义见图 5.4，引用该宏定义的例子见图 5.5。请注意，这里 mac.lib 是一个宏库，任何一个源程序文件中要引用其中某个宏只须在文件头部写上 include mac.lib 即可，当然，在使用 while 或 iff 之前，应先引用初始化宏 init_macs。

```
;*****************************
; The macros for label producing
; ——— using stack to store labels
;*****************************

connect macro x1,x2,x3,x4,x5
ifb <x3>
```

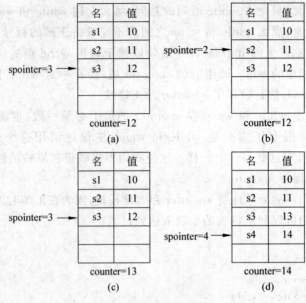

图 5.3 标号线

```
       &.x1&.x2   &.x4&.x5
else
   &.x1&.x2&.x3&.x4&.x5
endif
endm

labeling macro top
    connect ? _?,%s&.top,:,
        ; if top=1, then %s&.top will be the value of %s1
        ; . if %s1=002,then 'labeling top' will
        ; produce a label like '?_? 002:'.
endm

branch macro top,con
    connect j,con,,?_?,%s&.top
endm

st_asgn macro top,value
    connect s&.top,=,value,,
        ; produce a equation like 's1=002'.
endm

st_push macro
    spointer=spointer+1
    counter=counter+1
    st_asgn %spointer,%counter
        ; A sequence of equations like 's1=002'
        ; makes up a stack.
endm
```

```
st_pop macro
    spointer=spointer-1
endm

compares macro x1,op,x2
    ifidn <x1>,<ax>
    cmp x1,x2
    else
        ifidn <x2>,<ax>
    cmp x2,x1
        else
                    ; if none of x1 and x2 be AX,
                    ; then produce the next instructions.
    push ax
    mov ax,x1
    cmp ax,x2
    pop ax
        endif
    endif
endm

while macro x1,op,x2
local next
    st_push
    st_push
    sptmp=spointer-1
    labeling %sptmp            ; produce the loop head label.
    compares x1,op,x2
    connect j,op,,next
    branch %spointer,mp        ; jump out of the loop.
next:
endm

wend macro
    sptmp=spointer-1
    branch %sptmp,mp           ; jump to the beginning of the loop.
    labeling %spointer         ; produce the loop exit label.
    st_pop
    st_pop
endm

iff macro x1,op,x2
    local nxt
    st_push
    sptmp=spointer
    st_push
    compares   x1,op,x2
    connect  j,op,,nxt         ;to if clause
    branch   %sptmp,mp         ;branch to else clause
endm

elsee macro
branch %spointer,mp            ;to the end of whole if .
```

```
        labeling %sptmp              ;produce the label of else clause.
    endm

    ifend macro
        labeling %spointer
        st_pop
        sy_pop
    endm

    init_macs macro
        spointer=1
        counter=100
    endm
```

图 5.4 宏库 mac.lib

```
    include   mac.lib
    data segment
        char="y"
        array dw "y"
        char=char-1
        rept 19
            dw    char
            char=char-1
        endm
        dw " $ "
    data ends
    code segment
    assume ds:data,cs:code
    main proc far
    start:
        push ds
        sub ax,ax
        push ax
        mov ax,data
        mov ds,ax

        init_macs
        mov bx,0
        while bx,l,40
            mov ax,array[bx]
            mov bp,bx
            sub bp,2
            while ds:array[bp],g,ax
                push ds:array[bp]
                pop ds:array[bp+2]
                sub bp,2
                cmp bp,0
                jl   exit_while
            wend
    exit_while:
            add bx,2
        wend
```

```
        mov bx,0
        while bx,1,40
          mov dl,byte ptr array[bx]
          mov ah,2
          int 21h
          add bx,2
        wend
        ret
main endp
code ends
end start
```

图 5.5 使用 while 的一个例子

```
C   include mac.lib
C
C
C   ;****************************
C   ; The macros for label producing
C   ; --- using stack to store labels
C   ;****************************
C
C   connect macro x1,x2,x3,x4,x5
C   ifb <x3>
C       &x1&x2    &x4&x5
C   else
C       &x1&x2&x3&x4&x5
C   endif
C   endm
C
C   labeling macro top
C       connect ?_?,%s&top,:,
C   endm
C
C   branch macro top,con
C       connect j,con,,?_?,%s&top
C   endm
C
C   st_asgn macro top,value
C       connect s&top,=,value,,
C   endm
C
C   st_push macro
C       spointer=spointer+1
C       counter=counter+1
C       st_asgn %spointer,%counter
C   endm
C
C   st_pop macro
C       spointer=spointer-1
C   endm
C
C   compares macro x1,op,x2
C       ifidn <x1>,<ax>
```

```
C       cmp  x1,x2
C       else
C         ifidn <x2>,<ax>
C       cmp  x2,x1
C         else
C       push   ax
C       mov  ax,x1
C       cmp  ax,x2
C       pop  ax
C           endif
C       endif
C  endm
C
C  while   macro   x1,op,x2
C  local   next
C      st_ push
C      st_ push
C      sptmp=spointer-1
C      labeling %sptmp       ; produce the loop head label.
C      compares x1,op,x2
C      connect j,op,,next
C      branch %spointer,mp   ; jump out of the loop.
C  next:
C  endm
C
C  wend macro
C      sptmp=spointer-1
C      branch %sptmp,mp      ; jump to the beginning of the loop.
C      labeling %spointer    ; produce the loop exit label.
C      st_ pop
C      st_ pop
C  endm
C
C  iff   macro x1,op,x2
C      local nxt
C      st_ push
C      sptmp=spointer
C      st_ push
C      compares x1,op,x2
C      connect j,op,,nxt     ;to if clause
C      branch %sptmp,mp      ;branch to else clause
C  endm
C
C  elsee macro
C      branch %spointer,mp   ;to the end of whole if.
C      labeling %sptmp       ;produce the label of else clause.
```

```
C       endm
C
C       ifend macro
C           labeling %spointer
C           st_pop
C           sy_pop
C       endm
C
C       init_macs macro
C           spointer=1
C           counter=100
C       endm
                                    C
                                    C
                                    C
0000                                        data segment
 = 0079                                         char="y"
0000 0079                                       array dw "y"
 = 0078                                         char=char-1
                                                rept 19
                                                    dw char
                                                    char =char-1
                                                endm
0002  0078              +                       dw      char
0004  0077              +                       dw      char
0006  0076              +                       dw      char
0008  0075              +                       dw      char
000A  0074              +                       dw      char
000C  0073              +                       dw      char
000E  0072              +                       dw      char
0010  0071              +                       dw      char
0012  0070              +                       dw      char
0014  006F              +                       dw      char
0016  006E              +                       dw      char
0018  006D              +                       dw      char
001A  006C              +                       dw      char
001C  006B              +                       dw      char
001E  006A              +                       dw      char
0020  0069              +                       dw      char
0022  0068              +                       dw      char
0024  0067              +                       dw      char
0026  0066              +                       dw      char
0028  0024                                      dw      "$"
002A                                        data ends

0000                                        code segment
                                            assume   ds:data,cs:code
0000                                        main proc far
0000                                        start:
0000  1E                                        push ds
0001  2B C0                                     sub ax,ax
0003  50                              push    ax
0004  B8 ---- R                       mov ax,data
```

```
0007   8E D8                        mov ds,ax

                                    init_macs
0009   BB 0000                      mov bx,0
                                    while bx,l,40
000C                      +     ?_? 101:
000C   50                 +         push ax
000D   8B C3              +         mov ax,bx
000F   3D 0028            +         cmp ax,40
0012   58                 +         pop ax
0013   7C 03              +         jl ?? 0000
0015   EB 2D 90           +         jmp ?_? 102
0018                      +     ?? 0000:
0018   8B 87 0000 R                     mov ax,array[bx]
001C   8B EB                            mov bp,bx
001E   83 ED 02                         sub bp,2
                                        while ds:array[bp],g,ax
0021                      +?    _? 103:
0021   3E: 3B 86 0000 R   +             cmp    ax,ds:array[bp]
0026   7F 03              +             jg   ?? 0001
0028   EB 15 90           +             jmp  ?_? 104
002B                      +         ?? 0001:
002B   3E: FF B6 0000 R                     push ds:array[bp]
0030   3E: 8F 86 0002 R                     pop ds:array[bp+2]
0035   83 ED 02                             sub bp,2
0038   83 FD 00                             cmp bp,0
003B   7C 02                                jl exit_while
                                        wend
003D   EB E2              +             jmp ?_? 103
003F                      +     ?_? 104:
003F                                exit_while:
003F   83 C3 02                         add bx,2
                                    wend
0042   EB C8              +         jmp ?_? 101
0044                      +     ?_? 102:
0044   BB 0000                      mov bx,0
                                    while bx,l,40
0047                      +     ?_? 105:
0047   50                 +         push ax
0048   8B C3              +         mov ax,bx
004A   3D 0028            +         cmp ax,40
004D   58                 +         pop ax
004E   7C 03              +         jl ?? 0002
0050   EB 0E 90           +         jmp ?_? 106
0053                      +     ?? 0002:
0053   8A 97 0000 R                     mov dl,byte ptr array[bx]
0057   B4 02                            mov ah,2
0059   CD 21                            int 21h
005B   83 C3 02                         add bx,2
                                    wend
005E   EB E7              +         jmp ?_? 105
0060                      +     ?_? 106:
0060   CB                           ret
```

```
         0061                    main endp
         0061                    code ends
                                 end start
```

图 5.6 ex_51.asm 的列表文件

二、实验题

实验 5.1* 扩展 if 和 while 条件表达功能

1. 题目：

用宏和高级汇编技术实现高级语言中的 if 和 while 语句功能。

2. 要求：

在例 5.1 的宏库基础上扩充 iff 和 while 的功能,使其

(1) 不必限制条件表达式中关系操作数目,即允许用 and、or 和 not 连接多个关系操作,具体形式为:

```
    while   x,ge,AX,and,y,l,31
    if      not,x,l,y,or,BX,g,y
```

它们的高级语言形式为

```
while   (x>=AX)and(y<31)
if      not(x<y)or(BX>y)
```

(2) 条件表达式中没有括号的概念。

(3) 条件表达式中操作数除立即数外必须是字变量或字寄存器。

(4) 条件表达式必须在一行中写完,最多由 4 个关系操作构成,即最多有 3 个 and/or 和 4 个 not。

3. 提示：

在 if 和 while 的宏中,应设置足够的参数,即按可能出现的最多参数个数定义形式参数。另外,not 的出现与否不能确定,即预先给 not 留出的形式参数不一定对应实际参数时,在编写宏定义时应注意这一点。

5.2 连接技术

一、示例

例 5.2 可回卷的页显示 ex_52

一个程序可以由多个文件构成。在例 4.2 中,我们给出了一个程序,具有删除和改变页大小的分页显示能力,通过空格、P 和 D 三个命令来控制。现在我们希望为该程序再扩充一个功能——回卷,即显示完当前页后,如果用户键入 B 字符,将在屏幕上显示当前页的前一页,也就是回卷一页,继续分页显示。

原例 4.2 的程序已相当长,我们不希望将新增加的功能编入原例 4.2 的源文件中,可

以将该功能单独作为一个文件,让例 4.2 的主程序调用该文件中的过程,利用多模块的连接技术使它们正确连在一起执行。

为实现回卷,我们必须知道回卷一页相当于回卷多少个字符,据此计算出要显示的文件应将当前文件指针回移多少,在当前文件指针回移后,从它所指之处再读出一页来显示。这是回卷的基本思路。当然,为支持删除动作,临时文件中的当前文件指针也要同时移动。

为记录一页有多少个字符,设置了一个数组,该数组用于记录已显示过的每一行各包含多少字符,这样,在用户使用 P 命令改变过页大小后,仍然可以正确地确定回卷一页相当于回卷多少个字符。相应地,设置了一个当前所在行指针,以配合使用该数组。该数组和当前行指针分别叫 lines 和 cur_line。

lines 数组每个元素是某一行的字符数,该字符个数的登记由主程序(例 4.2)来完成,在每显示一个字符时,lines 的第 cur_line 个元素加 1,直至遇到回车换行时,使 cur_line 加 2(lines 是字数组),指向下一行。图 5.7,它是修改后的主程序框图。

具体的回卷动作与主程序分作两个源程序文件,文件名分别是 ex_52_1.asm 和 ex_52.asm,这两个文件有一个相同的数据段(data 段),由于 data 段定义为 common,这使得它们在连接后成为一个公共段,因而文件 ex_52.asm 和 ex_52_1.asm 中的程序可以共享该段中的所有变量。

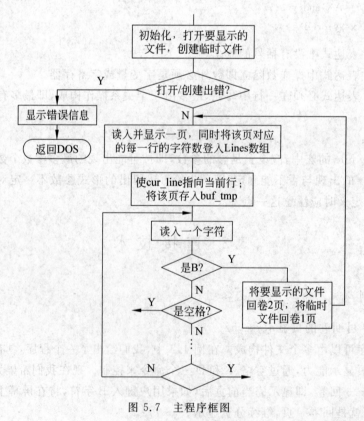

图 5.7 主程序框图

回卷动作由两部分组成:计算回卷距离和文件指针回移,参见图5.8的框图。图5.9和图5.10分别是主程序和回卷部分的源程序,后跟3个分号的语句为在原例4.2基础上增加或修改的语句,图5.11则是一个运行实例。

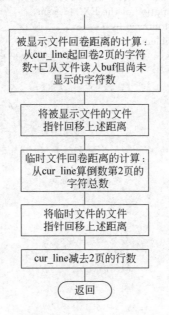

图 5.8 回卷动作流程

```
        extrn    rollback:far
        data segment common
            lines      dw    500 dup(?)
            cur_line   dw    0
            Pgsize     dw    ?
            buf_size   db    80
            s_buf      db    ?
            buf        db    200 dup(?)
            names      db    20 dup(?)
            cur        dw    ?
            handle     dw    ?
            buf_tmp    db    24*80 dup(?)
            cur_tmp    dw    ?
            name_tmp   db    "t0m1p",0
            handle_tmp dw    ?
            mark       db    ?

            mess_getname  db  0dh,0ah,"  Please input filename:$ "
            mess_err1     db  0ah,0dh,"  Illegal filename ! $ "
            mess_err2     db  0ah,0dh,"  File not found ! $ "
            mess_err3     db  0ah,0dh,"  File read error ! $ "
            mess_psize    db  0ah,0dh,"  Page Size : $ "
            mess_dele     db  0dh,0ah,"  The last page is deleted !"
            crlf          db  0ah,0dh," $ "
            mess_star     db  0ah,0dh," * * * * * * * * * * * * * * * * * "
```

```
                db      0ah,0dh,"$"
data ends

;* * * * * * * * * * * * * * * * * * * * * * * * * * * * * *
;*  This program is an augmentation of the EX_43.asm,       *
;*  all the instructions that is followed by ;;; are        *
;*  specially added for this problem.                       *
;* * * * * * * * * * * * * * * * * * * * * * * * * * * * * *

code segment
        assume ds:data, cs:code
        main proc far
start:
        push    ds
        sub     ax,ax
        push    ax
        mov     ax,data
        mov     ds,ax

        mov     cur_line,0              ;;;
        mov     bx,0                    ;;;
        mov     lines[bx],0             ;;;

        mov     mark,0
        mov     Pgsize,12
        mov     cur,200                 ; File data buffer is empty
        call    getline                 ; Get file name
        call    openf   ; open the file and creat temporary file,
                        ; (ax)=0 means no such file
        or      ax,ax
        jnz     display
        mov     dx,offset mess_err2
        mov     ah,09h
        int     21h
        jmp     file_end
display:
        mov     cx,Pgsize
        mov     cur_tmp,0
show_page:
        call    read_block ; read a line from handle to buf
        or      ax,ax
        jnz     next2
        mov     dx,offset mess_err3
        mov     ah,09h
        int     21h
        jmp     file_end
next2:
        call    show_and_reserve        ; display a line in buf,
                                        ; and put the line in buf_tmp.
                                        ; (bx) returned = 0
                                        ; means that the file reach its end.
        or      bx,bx
        jz      file_end
```

```
        or      cx,cx
        jnz     show_page

        mov     dx,offset mess_star
        mov     ah,09h
        int     21h
```

; the current page has been on screen, and followed by a line of stars.

```
wait_space:
    mov ah,1
    int 21h
    cmp al," "
    jnz psize
    call write_buf_tmp
    jmp display
psize:
    cmp al,"p"
    jnz back_page
    call write_buf_tmp
    call change_psize
    jmp stick
back_page:                          ;;;
    cmp al,"b"                      ;;;
    jnz delete                      ;;;
    call rollback                   ;;;
    cmp dx,0                        ;;;
    jz  stick                       ;;;
    jmp display                     ;;;
delete:
    cmp al,"d"
    jnz wait_space
    mov bx,Pgsize
    sub cur_line,bx
    mov mark,1
    mov dx,offset mess_dele
    mov ah,09h
    int 21h
stick:
    mov ah,1
    int 21h
    cmp al," "
    jnz stick
    jmp display
file_end:
    call write_buf_tmp
    cmp mark,0
    jz  ok
    call write_tmp_back
ok:
    ret
main endp
;*************************
```

```asm
;************************************
change_psize proc near
    push ax
    push bx
    push cx
    push dx
    mov  dx,offset mess_psize
    mov  ah,09h
    int  21h
    mov  ah,01
    int  21h

    cmp  al,0dh
    jz   illeg
    sub  al,"0"
    mov  cl,al
getp:
    mov  ah,1
    int  21h
    cmp  al,0dh
    jz   pgot
    sub  al,"0"
    mov  dl,al
    mov  al,cl
    mov  cl,dl              ; exchange al and cl.

    mov  bl,10
    mul  bl
    add  cl,al
    jmp  getp
pgot:
    mov  dl,0ah
    mov  ah,2
    int  21h                ; output 0ah to complete the RETURN.
    cmp  cx,0
    jle  illeg
    cmp  cx,24
    jg   illeg
    mov  PgSize,cx
illeg:
    mov  dl,0ah
    mov  ah,09h
    int  21h                ; output 0ah to complete the RETURN.
    pop  dx
    pop  cx
    pop  bx
    pop  ax
    ret
change_psize endp
;************************************

;************************************
```

```
openf proc near
    push    bx
    push    cx
    push    dx
    mov     dx,offset names
    mov     al,2
    mov     ah,3dh
    int     21h
    mov     handle,ax
    mov     ax,0
    jc      quit
    mov     dx,offset name_tmp
    mov     cx,0
    mov     ah,3ch
    int     21h
    mov     handle_tmp,ax
    jc      quit
    mov     ax,1
quit:
    pop     dx
    pop     cx
    pop     bx
    ret
openf endp
;* * * * * * * * * * * * * * * * * * * * * * * * * * *

;* * * * * * * * * * * * * * * * * * * * * * * * * * *
getline proc near
    push    ax
    push    bx
    push    cx
    push    dx
    mov     dx,offset mess_getname
    mov     ah,09h
    int     21h
    mov     dx,offset buf_size
    mov     ah,0ah
    int     21h
    mov     dx, offset crlf
    mov     ah,09h
    int     21h
    mov     bl,s_buf
    mov     bh,0
    mov     names[bx],0         ; insert 0 to form the asciiz string.
name_move:
    dec     bx
    mov     al,buf[bx]
    mov     names[bx],al        ; move the line got into name string
    jnz     name_move
    pop     dx
    pop     cx
    pop     bx
    pop     ax
```

```
        ret
getline endp
;* * * * * * * * * * * * * * * * * * * * * * * *


;* * * * * * * * * * * * * * * * * * * * * * * *
read_block proc near
        push    bx
        push    cx
        push    dx
        mov     ax,1
        cmp     cur,200
        jnz     back
        mov     cx,200
        mov     bx,handle
        mov     dx, offset buf
        mov     ah, 3fh
        int     21h
        mov     cur,0
        mov     ax,1
        jnc     back
        mov     cur,200
        mov     ax,0
back:
        pop     dx
        pop     cx
        pop     bx
        ret
read_block endp
;* * * * * * * * * * * * * * * * * * * * * * * *


;* * * * * * * * * * * * * * * * * * * * * * * *
show_and_reserve proc near
        push    ax
        push    dx
        mov     bx,cur
        mov     bp,cur_tmp
loop1:
        cmp     bx,200
        jl      lp
        jmp     exit
lp:
        mov     dl,buf[bx]
        mov     ds:buf_tmp[bp],dl
        inc     bx
        inc     cur
        inc     bp
        inc     cur_tmp
        mov     si,cur_line         ;;;
        inc     lines[si]           ;;;
        cmp     dl,1ah              ; search the file end
        jz      exit_eof
        mov     ah,02
        int     21h
```

```
            cmp     dl,0ah
            jz      exit_ln
            jmp     loop1
exit_eof:
            mov     bx,0
            jmp     exit
exit_ln:
            add     cur_line,2          ;;;
            mov     bx,cur_line         ;;;
            mov     lines[bx],0         ;;;
            dec     cx
exit:
            pop     dx
            pop     ax
            ret
show_and_reserve endp
;******************************


;******************************
write_buf_tmp proc near
            push    ax
            push    bx
            push    cx
            push    dx
            mov     dx,offset buf_tmp
            mov     cx,cur_tmp
            mov     bx,handle_tmp
            mov     ah,40h
            int     21h
            pop     dx
            pop     cx
            pop     bx
            pop     ax
            ret
write_buf_tmp endp
;******************************


;******************************
write_tmp_back proc near
            push    ax
            push    bx
            push    cx
            push    dx
            mov     bx,handle_tmp
            mov     ah,3eh
            int     21h                 ; close the temporary file.
            mov     bx,handle
            mov     ah,3eh
            int     21h                 ; close the file giving.
            mov     dx,offset name_tmp
            mov     al,0
            mov     ah,3dh
            int     21h                 ; open the temporary file for reading.
```

```
        mov     handle_tmp,ax
        mov dx,offset names
        mov al,1
        mov ah,3dh
        int     21h                 ; reopen the file giving for writing.
        mov     handle,ax
        mov     si,1
wrt_back:
        mov bx,handle_tmp
        mov ah,3fh
        mov cx,200
        mov dx,offset buf
        int     21h                 ; read a page bytes from temporary file to buf.
        jc  wrt_end
        mov si,ax
        mov bx,handle
        mov ah,40h
        mov cx,200
        mov dx,offset buf
        int     21h                 ; write a page bytes from buf to the file giving.
        or      si,si
        jnz wrt_back
        mov ah,3eh
        mov bx,handle
        int     21h                 ; close the file giving.
wrt_end:
        pop dx
        pop cx
        pop bx
        pop ax
        ret
write_tmp_back endp
;************************
code ends
end start
```

图 5.9　主程序源程序

```
public rollback
data segment common
    lines    dw   500 dup(?)
    cur_line dw 0
    Pgsize dw ?
    buf_size db   80
    s_buf    db ?
    buf      db   200 dup(?)
    names    db   20 dup(?)
    cur      dw ?
    handle   dw ?
    buf_tmp  db   24*80 dup(?)
    cur_tmp  dw ?
    name_tmp db   "t0m1p",0
    handle_tmp dw ?
    mark     db ?
```

```
            mess_getname   db   0dh,0ah,"   Please input filename: $"
            mess_err1      db   0ah,0dh,"   Illegal filename ! $"
            mess_err2      db   0ah,0dh,"   File not found ! $"
            mess_err3      db   0ah,0dh,"   File read error ! $"
            mess_psize     db   0ah,0dh,"   Page Size : $"
            mess_dele      db   0dh,0ah,"   The last page is deleted !"
            crlf           db   0ah,0dh,"$"
            mess_star      db   0ah,0dh,"* * * * * * * * * * * * * * * * * * * * * * *"
                           db   0ah,0dh,"$"
data ends

code1 segment
        assume cs:code1,ds:data
        rollback proc far

            mov     bx,Pgsize
            shl     bx,1
            mov     cx,bx                   ; need to back double page for the source file.
            call    count_char              ;count the number of chars need to back (dx).

            cmp     dx,0
            jne     next
            ret
next:
            mov     bx,200
            sub     bx,cur
            add     dx,bx                   ; Double page + chars not displayed in buf to back.
            mov     bx,handle
            call    movefilepointer

            mov     cx,Pgsize
            sub     cur_line,cx
            jle     ok
            call    count_char              ; the temp file move back one page, because
                                            ; the last page in buf_tmp have not been
                                            ; writen into the temp file.
            mov     bx,handle_tmp
            call    movefilepointer

            mov     cx,Pgsize
            sub     cur_line,cx
            jge     ok
            mov     cur_line,0
ok:
            mov     cur,200
            mov     mark,0
            mov     cur_tmp,0

            ret
        rollback endp

        count_char proc near
            mov     dx,0
```

```
            mov     bx,cur_line
again:
            sub     bx,2
            jl      ok
            add     dx,lines[bx]
            loop    again
ok:
            ret
count_char endp

movefilepointer proc near
            push    dx
            neg     dx
            mov     cx,0
            not     cx
            mov     ah,42h
            mov     al,1
            int     21h
            pop     dx
            ret
movefilepointer endp

code1 ends
            end
```

图 5.10 回卷动作源程序

Please input filename：mac.lib

```
;*******************************
; The macros for label producing
; --- using stack to store labels
;*******************************

connect macro x1,x2,x3,x4,x5
ifb <x3>
    &x1&x2       &x4&x5
else
&x1&x2&x3&x4&x5

*******************************
endif
endm

labeling macro top
    connect ?_?,%s&top,:,
endm

branch macro top,con
    connect j,con,,?_?,%s&top
endm

st_asgn macro top,value
```

· 170 ·

```
* * * * * * * * * * * * * * * * * * * * * * * * * * * * * *
    connect s&top,=,value,,
endm

st_push macro
    spointer=spointer+1
    counter=counter+1
    st_asgn %spointer,%counter
endm

st_pop macro
    spointer=spointer-1
endm

* * * * * * * * * * * * * * * * * * * * * * * * * * * * * *
p
    Page Size : 6

compares macro x1,op,x2
    ifidn <x1>,<ax>
    cmp x1,x2
    else
       ifidn <x2>,<ax>
* * * * * * * * * * * * * * * * * * * * * * * * * * * * * *
b    st_asgn %spointer,%counter
endm

st_pop macro
    spointer=spointer-1
endm

* * * * * * * * * * * * * * * * * * * * * * * * * * * * * *

compares macro x1,op,x2
    ifidn <x1>,<ax>
    cmp x1,x2
    else
       ifidn <x2>,<ax>

* * * * * * * * * * * * * * * * * * * * * * * * * * * * * *
    cmp x2,x1
       else
    push ax
    mov ax,x1
    cmp ax,x2
    pop ax

* * * * * * * * * * * * * * * * * * * * * * * * * * * * * *
       endif
    endif
endm

while macro x1,op,x2
```

```
        local next

* * * * * * * * * * * * * * * * * * * * * * * * * * * * *
b       cmp x2,x1
          else
        push ax
        mov ax,x1
        cmp ax,x2
        pop ax


* * * * * * * * * * * * * * * * * * * * * * * * * * * * *
            endif
          endif
endm

while macro x1,op,x2
local next
* * * * * * * * * * * * * * * * * * * * * * * * * * * * *
        st_ push
        st_ push
        sptmp=spointer-1
        labeling %sptmp          ; produce the loop head label.
        compares x1,op,x2
        connect j,op,,next

* * * * * * * * * * * * * * * * * * * * * * * * * * * * *
p
        Page Size : 12
          branch %spointer,mp ; jump out of the loop.
next:
endm

wend macro
        sptmp=spointer-1
        branch %sptmp,mp
            ; jump to the beginning of the loop.
        labeling %spointer
            ; produce the loop exit label.
        st_ pop
        st_ pop
endm


* * * * * * * * * * * * * * * * * * * * * * * * * * * * *
iff    macro x1,op,x2
        local nxt
        st_ push
        sptmp=spointer
        st_ push
        compares    x1,op,x2
        connect     j,op,,nxt ;to if clause
        branch      %sptmp,mp ;branch to else clause
endm
```

```
elsee macro
    branch %spointer,mp ;to the end of whole if .
* * * * * * * * * * * * * * * * * * * * * * * * * * * * *
        labeling %sptmp
            ;produce the label of else clause.
endm

ifend macro
    labeling %spointer
    st_ pop
    sy_ pop
endm

init_ macs macro
    spointer=1
    counter=100


* * * * * * * * * * * * * * * * * * * * * * * * * * * * *
endm
```

图 5.11 运行实例

请注意一下回卷动作中显示文件和临时文件的回卷字符数不一样,这是因为已显示的最后一页在键入"B"时尚未被写入临时文件,这时临时文件和显示文件的当前文件指针位置并不相同,故此回卷字符数的计算也不一样。

二、实验题

实验 5.2 菜单使用

1.题目: 菜单

2.要求:

用菜单方式将已做过的所有程序组织起来,以便随意运行某一程序或查阅其源程序,具体要求如下:

(1)分两级菜单:主菜单列出所有程序名;二级菜单则列出该菜单程序的两项功能:运行指定程序或显示指定程序源文件。

(2)用户指定主菜单中一程序后自动切换到二级菜单,切换前要清屏。

(3)两级菜单中均包含一退出选择项,当且仅当用户选择该项时退出当前菜单,返回上一级菜单(对主菜单而言是返回到 DOS),当然返回前要清屏。

(4)每个选择项前有个数字,用户键入相应数字表示选择了相应的项。

(5)显示源程序时应以分页方式进行。

(6)在程序刚执行,还未进入主菜单前,应显示一个说明性菜单,用英文简单介绍一下你的菜单程序功能和用法,同时用扬声器奏出一段音乐,音乐一结束,立即显示主菜单。

3.提示:

该菜单程序主要涉及到以下几个问题:多个程序如何连接在一起;如何显示菜单;如何奏乐;如何显示源程序;如何执行一个程序。这里面每个问题我们都不难解决,大多是已做过的实验,只要能正确地将它们组织起来,该菜单程序就很容易实现了。

　　用户在选择菜单中某一项时是输入数字的方式,因而用表的方式来寻找相应的源程序文件名或程序起始地址可能是一个很方便的办法,对应到数据段的具体设计就很容易了。比如,我们可以设置如下两个数据结构:

```
source_file1  db  'abc.asm'
source_file2  db  'efg.asm'
    ...
source_filen  db  'xyz.asm'
file_table    dw  source_file1
              dw  source_file2
              ...
              dw  source_filen

exec_table    dw  program1
              dw  program2
              ...
              dw  programn
```

　　通过 file_table 数组和用户键入的数字,很容易获得相应的源文件名。当然,最好将源文件名字符串按文件名的标准格式放好,这会给文件的打开带来方便。至于 exec_table 数组是一个跳跃表,用间接转移方式可转去 programi 处执行相应程序段。

附录一　上机基本操作

1. 开机和关机

IBM-PC 兼容机的电源开关在机器的右侧。但打开机器电源前,首先将显示器的最上一个旋扭转至 0,使该扭上方一个小指示灯亮。然后再打开机器右侧的电源开关,等待显示屏幕出现提示符 C>,当出现 C>后方可插入软盘。

在关机前请你一定要退出到 C>状态后再关掉电源开关,将显示器上的旋扭转至 1。

2. 汇编语言程序的上机过程

使用汇编语言程序上机需经过四个步骤:

(1) 调用全屏幕编辑程序 WordStar 或 PCED 或 EDLTN,建立和修改源程序;

(2) 将源程序经过汇编,变成机器代码形式的目标文件(OBJ);

(3) 经过连接程序处理,形成可执行文件(EXE);

(4) 利用 DEBUG 调试程序单步执行或利用设断点的方法运行 EXE 程序,检查程序中的错误。

以上四步,由下图简要说明:

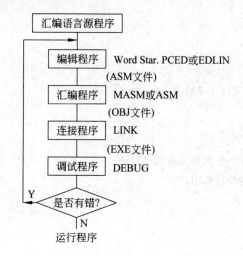

3. 常用 DOS 命令

(1) 查看目录命令 DIR,它列出所指盘上的文件目录,如

　　C>DIR A:　　或　　C>DIR A:/W

它们将列出 A 盘上全部文件。

(2) 显示命令 TYPE,它将磁盘上所指文件的内容显示在屏幕上或在打印机上输出(若打印机已联机)。如:

　　C>TYPE TEST.ASM

此命令将文件 TEST.ASM 的内容显示出来。

(3) 拷贝命令 COPY,它把一个或多个文件拷贝成副本,如:

A>COPY EX1.EXE B:

175

将把 A 盘的文件 EX1. EXE 拷贝到 B 盘上。

(4) 改名命令 RENAME，如

A＞RENAME EX1. EXE EX2. EXE

将把 A 盘上文件 EX1. EXE 改名为 EX2. EXE。

(5) 删除命令 ERASE 或 DEL，它将从指定的驱动器上删除一个或多个文件。如

C＞ERASE A：EX1. ASM

将删除驱动器 A 上文件 EX1. ASM。

(6) 磁盘格式化命令 FORMAT，如

C＞FORMAT A：

将为驱动器 A 上的软盘格式化（注意：一般不允许格式化硬盘）。

4. 常用目录操作命令

(1) 建立子目录，格式如下：

C＞ MD　子目录名

或

C＞ MKDIR　子目录名

例如，

C＞ MD　　C:\sales

则在硬盘 C 的根目录下建立了一个子目录 sales。

又如：

C＞ MD　　C:\sales\wan

则在子目录 sales 下再建立了一个名为 wan 的子目录。

上面两个例子中的"\"为路径标志。

(2)　删除子目录，格式为

C＞RD　子目录名

或

C＞RMDIR　子目录名

例如，

C＞ RD　　C:\sales\wan

则把子目录 wan 删除，条件是子目录 wan 必须是空的，且 wan 不能是当前目录。

(3)　改变当前目录，格式为：

C＞CD　目录名

或

C＞CMDIR　目录名

将与目录名对应的子目录变为当前目录。

(4) 显示目录结构,格式为：

C>Tree

将产生一个描述 C 盘整个目录的报告。

(5) 退出子目录,格式为：

C>CD..

将退出当前子目录到直接外层目录下。

附录二 全屏幕编辑程序 WordStar

WordStar 是一个文字处理软件,可以用它编辑汇编语言程序,也可以用来拟稿,写文件等。它具有分段、合并段、移动段、插入、删除、查找、修改内容等功能,除此之外,还可以用它进行拷贝、删除文件等操作。

1. 启动 WordStar

键入 WS,即形如:

C>WS

将启动 WordStar 程序(设盘 C 中已存有 WordStar 软件),屏幕上出现"起始命令表"如下:

not editing

《OPENING MENU》

——Preliminery Commands——	——File Commands——	——System Commands——
L Change logged disk drive		R Run a program
F File directory now ON	P Print a file	X EXIT to system
H Set help level	E RENAME a file	
——Commands to open a file-		-WordStar Options-
D Open a Document file	O COPY a file	M Run Mail Merge
N Open an Non-document file	Y DELETE a file	S Run SpellStar

其中:

L 改变当前磁盘;   E 更换文件名;
F 改变目录显示状态;   O 拷贝文件;
H 设置联机求助(HELP)详略程度; Y 删除文件;
D 编辑文本文件;   R 运行程序;
N 编辑非文本文件;   X 退出 WordStar。
P 打印/中断打印文件;

2. 编辑命令

(1) 帮助命令^J(^J 表示 Ctrl 键与 J 键同时按下)。

用来查看系统提供的帮助信息,以了解有关的命令注释。

(2) 退出编辑状态命令,分三种:

① ^KD(或 PF1 键)命令,它将当前被编辑的文件存入磁盘,然后退出编辑状态,返回到"起始命令表";

② ^KX 命令,它与^KD 相同,但最后返回到操作系统下;

③ ^KQ(或 PF2 键)命令,它放弃当前被编辑文件,退出编辑状态,返回到"起始命令表"。

3. 光标移动命令

↑ 或 ^E:光标上移一行

↓ 或 ^X:光标下移一行

← 或 ^S:光标左移一个字符

→ 或 ^D:光标右移一个字符

^A   ：光标左移一句（或一个英文单词）
^F   ：光标右移一句（或一个英文单词）
^R   ：光标移到当前显示页的前一页
^C   ：光标移到当前显示页的后一页
^W   ：光标不动而整个屏幕向下移一行
^Z   ：光标不动而整个屏幕向上移一行
^QE   ：光标移到屏首
^QX   ：光标移到屏末
^QR 或 PF9 键：光标移到文件头
^QC 或 PF10 键：光标移到文件尾
^QB   ：光标移到块首
^QE   ：光标移到块尾

4. 删除与插入命令

^G   ：删除光标处的字符
DEL 键   ：删除光标左边的字符
^T   ：删除光标右边一句
^Y   ：删除光标所在行
^V 或 INS 键：变换插入/非插入状态
^N   ：在光标处加一行

5. 移动、复制命令

(1) 在进行移动、复制等操作前，先把光标移到字块前后，分别加注标记，方法是：

^KB 或 PF7 键   ：在光标后加块首标记
^KK 或 PF8 键   ：在光标前加块尾标记
^KH   ：删除字块的首尾标记

再将光标移到预想位置，并发出命令：

^KV   ：将加有首尾标记的字块移到当前光标位置
^KC   ：将加有首尾标记的字块拷贝到当前光标位置，原字块仍存在
^KY   ：删除加有首尾标记的字块

(2) 文件之间的内容交换可用如下命令：

^KW   ：把当前文件中加有首尾标记的字块写入指定文件
^KR   ：把指定文件的内容拷贝到当前文件中光标所在位置

6. 打印命令

在"起始命令表"下，按 P 键，屏幕上提示输入被打印文件的名称，键入文件名后，屏幕上继续出现多个问题，如：是否输出到磁盘上(Y/N)？按 N 键或回车表示不输出到磁盘上。随后又提问从第几页打印到第几页等等，回答完全部问题即可打印。如果需要打印整个文件，可在输入文件名后，按 ESC 键，则直接进入打印。在打印期间，随时按^KP可以暂停或继续打印工作。

附录三 全屏幕编辑程序 pced

1. 进入和退出 pced

当你开机以后插入 A 盘,进入 A>状态下,键入 pced 空一格后输入你的文件名和扩展名,一回车就进入编辑状态,具体格式如下:

　　　　A>pced　文件名.asm\

这时可以输入你的源程序。当你的源程序输入完毕,要存盘,你按两次 F10 键即可将文件存盘并退出,如果你不存盘退出用 Shift 键帮忙,即同时按 Shift 和 F10 键。

2. 功能键的使用

F1：按此键,光标移到文件首,屏幕上显示从文件首开始的第一页内容。

F2：按此键,光标移到文件尾,屏幕上显示文件的最后一页。

F3：文件指针和光标移到原所在段的下一段首,段是以一个或多个空行为间隔的文件块,当无下一个文件段时指针移到文件首。

F4：完成一次搜索操作,被搜索的字符由 Shift+F4 键取得,同时,输入这一字符时也以 F4 键作为结束符。程序从光标指针位置开始向后搜索这一字符串,若到文件尾都没寻找到,程序继续从文件首到指针位置之间搜索。如仍找不到则响铃提示。若找到,指针就移到所找到的第一个字符串的字符串首。

F5：光标在本行中左移 40 列(或 16 列,半屏),若从光标开始到行首尚未达到要移动的列数,则光标移到行首。

F6：光标在本行内右移 40 列(或 16 列),若从光标到行尾少于这一列数,则光标移到行尾。

F7：光标移到指定的行,按 F7 后,提示行显示出要求输入行数的指示,用户在此后输入最多 5 个的 ASCII 码数字,少于 5 个用回车(Return)结束。输入为 0 时或空输入时,操作废除,光标回到原来位置。当数字输入完毕后,光标就移到从文件首开始算起的这一行上去(从 1 数起)。若行值超过了文件的行数,则移到文件尾。

F8：光标移到指定的列,列值也由提示行输入的 ASCII 数值得到,当输入的值过大时,光标移到行尾。

F9：连续按两次 F9 键,将编辑的文本内容存盘,存盘后,继续进行编辑操作。

该键的另一功能：当制表标志(Shift+F9)或设置范围标志(Shift+F7)有效时,这一键仅完成废除这些标志设置的操作。

F10：连续按两次 F10 键,将文本内容存盘,同时程序退出到操作系统。

3. Shift 键与功能键同时使用

Shift+F1：完成字符串替换功能,分两种情况：

　　　　1)　当未设定范围时(范围的定义见 Shift+F7 中说明),完成一次替换操作；被替换的字符串由 Shift+F4 键取得,和 F4 搜索功能使用的是同一个字符串,若光标位置开始的文件内容与这一字符串不同,这一键仅完成一次搜索功能；若光标位置开始的文件内容与这一字符串相同,将完成一次替换,这一字符串被清除,替入的字符串在后备缓冲区保存着,而这一缓冲区的内容由 Shift+F7 键获得。这一情况可完成选择替换的功能。

　　　　2)　当范围已设定,那么替换操作就在这一范围内执行。程序将这一范围内所有与上述相符的字符串替换为新的字符串,而范围以外的内容完全不变。替换完毕后,原来设置的范围及其状态撤除。这一情况可完成一组替换的功能。

Shift+F2:磁盘功能操作,分两种情况:

1) 当未设定范围时,这一键完成读文件操作。此时,在提示行提示要求输入文件的文件名,文件名输入完毕后(用 Return 结束),编辑程序将这一文件名的文件读入,并插入到当前光标位置。若输入的文件名为空,程序不进行读盘操作,光标回到原来位置,下面情况也类似。

2) 当已设定了范围后,这一键完成写文件操作。用户在提示行上键入所需写入的文件名,用 Return 键结束后,编辑程序将所设范围的内容写入盘上这一文件名的文件中,同时这一范围的内容被放入到后备缓冲区中。

Shift+F3:制表功能键

这是个双功能键,第一次按动此键,程序将当前光标位置记录下来,作为制表的起始点,再次键入 Shift+F3 作为制表标志的结束点,也是制表操作的真正开始键。程序根据上次记录的位置和第二次按下 Shift+F3 时光标位置绘制不同类型表格:

1) 两个位置在同一行上,而不在同一列时,绘制出一条横线;
2) 两个位置在同一列上(汉字显示方式下,列间距小于等于1),而不在同一行时,绘制出一条纵线;
3) 两个位置的行、列都不同时,绘制以这两点为对角顶点的矩形方框。

这里要说明几点:

1) 画线或画方框采取覆盖方式,线条经过之处,原有内容抹除,换上表格线,表格线未经过的其他位置内容不变,若表格线所占空间在文件内是空白(无内容)则程序自动补上空格符;
2) 画表格线仅在原来字符的基础上加线,也就是,若表格线所经过之处已为表格线符,程序仅仅将原表格线更换,在保留原表格线结构的基础上变为增加了所画之线的表格符;
3) 在汉字方式下,出现的是汉字表格符(占两个位置)在 ASCII 方式下为表格图形符,占一个位置。

Shift+F4:字符串取模键

按这一键后,在编辑提示行提示输入要求搜索的字符串,字符串可以输入任何字符,包括各种控制符。输入的字符显示在提示行,控制符全用其显示码代替。输入回车符等效于输入了回车,换行两个键。输入结束后,用 F4 结束字符串的输入,并开始第一次搜索。若输入的字符串为空(未输入一个字符或输入的字符已全部被删除)则缓冲区的原字符串仍有效,且此时按 F4 将不进行搜索操作。

输入的这一字符串对搜索与替换两项操作均有效。

Shift+F5:删行尾

此键的功能是将自光标位置开始到行尾的内容以及回车、换行符清除掉,把下行的内容连接到光标处。若本行为文件的最后一行,则删除直至文件尾的内容。

Shift+F6:删行首

此键完成删除从行首到光标位置(不包括光标位置)的内容,光标及其后的本行内容前移到行首。

Shift+F7:这一键是双功能键

第一次按下此键开始设置范围,当前光标位置就作为设置范围的起止点,以后字符输入和光标移动等多种操作就使得从这一起止点开始到光标位置（或从光标位置到这一起止点,取决于哪一点在前）的所有内容反相显示,以便于提示用户其所设定的范围区间。

第二次按下 Shift＋F7 键,就使上述起止点到当前光标位置之间的内容在文件和屏幕中清除掉,并将这一块内容移入后备缓冲区中,作为以后的恢复块功能（Shift＋F8）之用,也作为字符串替换时串插入（目的串）之用。

当范围设定后（第一次按下 Shift＋F7 后）,还有一些操作是基于这一范围之上的设定与大小之上工作的,它们是：

1) 字符串替换,其替换范围限于这一范围之内,参见 Shift＋F1 功能；
2) 磁盘操作,将设定的范围的内容写入特定的文件中,参见 Shift＋F2 功能；
3) 字处理,字处理也是基于这一范围规定的行之间,参见 Shift＋F9 功能。

Shift＋F8：块插入

按下这一键,编辑程序将后备缓冲区的内容（由 Shift＋F7 取得）插入到当前光标位置。这一功能可以是重复性的,这对于文件块的移动,文件块复制极其方便。

Shift＋F9：得到字处理的最大长度。

Shift＋F10：终止编辑返回操作系统。

Ctrl＋Q 与此功能相同。

4. 小键盘功能

(1) 在正常情况下,即 NumLock 键未按下或小键盘的键不是与 Shift 键同时按下时,

7 Home	8 ↑	9 PgUp	PrtSc
4 ←	5	6 →	−
1 End	2 ↓	3 PgDn	＋
0 Ins	· Del		

↑ [8]：光标上移一行；
↓ [2]：光标下移一行；
← [4]：光标左移一个位置；
→ [6]：光标右移一个位置
Home [7]：将光标移到本行行首；
End [1]：将光标移到本行行尾
PgUp [9]：光标上移一页；
PgDn [3]：光标下移一页；
Ins [0]：输入方式切换；改变编辑程序输入时的状态；使编辑输入状态在插入（Insert）和覆盖（Cover）两种方式下进行切换；
Del [.]：删除光标位置的一个字符；
[＋]：删字；

删除从光标位置到本行下一个字首(若无下个字,则到行尾)之间内容;当光标在行尾时,删除回车换行符;若光标在文件尾时,则不操作。

[→]:光标右移一个字;

光标右移到本行下一个字的字首,若本行下一个字不存在,光标则移到下一行行首;若这时已为文件尾行,则光标移到文件尾。

(2) 当 NumLock 键按下时,或小键盘的键与 Shift 键同时按下时,小键盘对应的则是表格符或特殊符号。

[1]：└　　　[8]：┬
[2]：⊥　　　[9]：┐
[3]：┘　　　[0]：。
[4]：├　　　[+]：│
[5]：┼　　　[-]：─
[6]：┤　　　[.]：、
[7]：┌

表格符在 ASCII 码方式与汉字方式时输入的代码是不同的。

5. 控制命令

CTRL+E　光标上移一行　　　CTRL+F　光标右移一字
CTRL+X　光标下移一行　　　CTRL+T　删除右边一字
CTRL+S　光标左移一列　　　CTRL+G　删除光标字符
CTRL+D　光标右移一列　　　CTRL+H　光标左移一列
CTRL+W　屏幕下滚一行　　　CTR+Y　删除光标一行
CTRL+Z　屏幕上滚一行　　　CTRL+M　输入回车代码
CTRL+R　屏幕上滚一页　　　CTRL+N　插入回车代码
CTRL+C　屏幕下滚一页　　　CTRL+I　输入 TAB 码
CTRL+A　光标左移一字　　　CTRL+V　插入/覆盖方式切换

附录四　行编辑程序 EDLIN

行编辑程序是按行方式对源程序进行输入、显示、修改的工具。

1. 行编辑程序的调用

在 DOS 提示符下键入命令 EDLIN 及文件名,如:

C＞ EDLIN　A:EX1.ASM

这时若指定文件名存在,则将显示:

　　END OF INPUT FILE
　　＊─

接着可输入 EDLIN 的各个命令(＊是 EDLIN 的提示符)。若指定文件不存在,则 EDLIN 建立一个新文件,并显示:

　　NEW FILE
　　＊─

此时可输入各种命令。

2. EDLIN 的主要命令

(1) 插入命令 I,格式为

　　[行号]　I

对于新文件,可在提示符＊后直接写 I。

退出插入方式是按 CTRL 和 BREAK 键。

(2) 显示行命令 L,格式为

　　[起始行号]　[,终止行号]　L

若某个行号省略,则按指定行号向前或向后显示 23 行,若两个行号都被省略,则从现行行的前 11 行开始共显示 23 行。

(3) 编辑一行的命令,格式为:

　　[行号] Enter 键

则在屏幕上首先显示指定的行号和该行内容,然后在下一行显示指定行号,等待输入编辑命令。

(4) 删除行命令 D,格式为

　　[行号]　[,行号] D

若省略一个行号,则从当前行至指定行被删除,若省略两个行号,则只删除当前行号。

(5) 搜索命令 S,格式为:

　　[行号]　[,行号]　[?]　S[字符串]

此命令在指定范围内搜索一个规定的字符串,显示该字符串所在行,将此行变为现行行。

(6) 替换命令 R,格式为:

[行号] [,行号] [?] R[字符串 1][<PF6> 字符串 2]

此命令将规定范围内的所有字符串 1,替换为字符 2 的形式。

(7) 读行命令 A,格式为:

[n] A

把指定数目(n)的行从盘上读到内存中被编辑的文件。

(8) 复制命令 C,格式为:

[行号 1],[行号 2],行号 3[,重复次数]

C 把行号 1 至行号 2 范围内的行复制到由行号 3 开始的行上,并重复复制指定的次数。

(9) 移动命令 M,格式为:

[行号 1][,行号 2],行号 3　M

把行号 1 至行号 2 范围内的行移到行号 3 的前面。

(10) 页显示命令 P,格式为:

[行号 1][,行号 2] P

把行号 1 至行号 2 范围内的行以页的形式显示出来,修改当前行为显示的最后一行。

(11) 读文件命令 T,格式为:

[行号] T [驱动器名:]　文件名

把指定文件的内容读入内存,插入到正在编辑的文件中指定行号之前。

(12) 写入命令 W,格式为:

[n] W

把正在编辑的文件从行号 1 开始连续写 n 行到软盘上去,剩下部分的行号重新编号。

3. 退出 EDLIN 命令

(1) E 命令,格式为:

　　＊E
　　C>

它把编辑缓冲区中的内容存盘,退出 EDLIN,返回 DOS。

(2) Q 命令,格式为:

　　＊Q
　　Abort edit (Y/N)? Y
　　C>

它不保留编辑缓冲区的内容,退出 EDLIN,返回 DOS。

附录五 调试程序 DEBUG

DEBUG 是专门为汇编语言设计的一种调试工具,它通过步进,设置断点等方式为汇编语言程序员提供了非常有效的调试手段。

1. DEBUG 程序的调用

在 DOS 提示符下,可键入命令:

C> DEBUG [d:][path][文件名][参数 1][参数 2]

其中文件名是被调试文件的名字,它须是执行文件(EXE),两个参数是运行被调试文件时所需要的命令参数,在 DEBUG 程序调入后,出现提示符"-",此时,可键入所需的 DEBUG 命令。

2. DEBUG 的主要命令

(1) 显示内存单元内容的命令 D,格式为:

　　　　-D[地址]

或

　　　　-D[范围]

(2) 修改内存单元内容的命令 E,它有两种格式

1) 用给定的内容代替指定范围的单元内容:

　　　　-E 地址　内容表

例如:

　　　　-E DS:100 F3 "XYZ" 8D

其中 F3, "X","Y","Z"和 8D 各占一个字节,用这五个字节代替原内存单元 DS:100 到 104 的内容, "X", "Y","Z"将分别按它们的 ASCII 码值代入。

2) 逐个单元相继地修改:

　　　　-E 地址

例如:

　　　　-E 100:

18E4:0100 89.78

此命令是将原 100 号单元的内容 89 改为 78。78 是程序员键入的。

(3) 检查和修改寄存器内容的命令 R,它有三种方式:

1) 显示 CPU 内部所有寄存器内容和标志位状态;格式为:

　　　　-R

R 命令显示中标志位状态的含义如下表所示:

标 志 名	置 位	复 位
溢出 Overflow(是/否)	OV	NV
方向 Direction(减量/增量)	DN	UP
中断 Interrupt(允许/屏蔽)	EI	DI
符号 Sign(负/正)	NG	PL
零 Zero(是/否)	ZR	NZ
辅助进位 Auxiliary Carry(是/否)	AC	NA
奇偶 Parity(偶/奇)	PE	PO
进位 Carry(是/否)	CY	NC

2) 显示和修改某个指定寄存器内容,格式为:

　　－R 寄存器名

　　例如打入:－R AX

　　系统将响应如下:

　　　　AX F1F4

　　　　:

表示 AX 当前内容为 F1F4,此时若不对其作修改,可按 ENTER 键,否则,打入修改后内容,如:

　　－R BX

　　BX 0369

　　:059F

则 BX 内容由 0369 改为 059F。

3) 显示和修改标志位状态,命令格式为:

　　－RF

　　系统将给出响应,如

　　OV DN EI NG ZR AC PE CY－

　　这时若不作修改可按 Enter 键,否则在"－"号之后键入修改值,键入顺序任意。如

　　OV DN EI NG ZR AC PE CY－PONZDINV

(4) 运行命令 G,格式为:

　　－G[＝地址 1][地址 2[地址 3...]]

　　其中地址 1 规定了运行起始地址,后面的若干地址均为断点地址。

(5) 追踪命令 T,有两种格式:

　　1) 逐条指令追踪:

　　　　－T[＝地址]

　　该命令从指定地址起执行一条指令后停下来,显示寄存器内容和状态值。

　　2) 多条指令追踪:

　　　　－T[＝地址][值]

　　该命令从指定地址起执行 n 条命令后停下来,n 由[值]确定。

(6) 汇编命令 A,格式为：

　　－A［地址］

　　该命令从指定地址开始允许输入汇编语句,把它们汇编成机器代码相继存放在从指定地址开始的存储器中。

(7) 反汇编命令 U,有两种格式：

　　1) －U［地址］

　　　该命令从指定地址开始,反汇编 32 个字节,若地址省略,则从上一个 U 命令的最后一条指令的下一单元开始显示 32 个字节。

　　2) －U 范围

　　　该命令对指定范围的内存单元进行反汇编,例如：

　　　　－U 04BA：0100 0108 或

　　　　－U 04BA：0100 L9

　　　此二命令是等效的。

(8) 命名命令 N,格式为：

　　－N 文件标识符［文件标识符］

　　此命令将两个文件标识符格式化在 CS：5CH 和 CS：6CH 的两个文件控制块内,以便使用 L 或 W 命令把文件装入或者存盘。

(9) 装入命令 L,它有两种功能：

　　1) 把磁盘上指定扇区的内容装入到内存指定地址起始的单元中,格式为：

　　　　－L　地址　驱动器　扇区号　扇区数

　　2) 装入指定文件,格式为：

　　　　－L［地址］

　　　此命令装入已在 CS：5CH 中格式化的文件控制块所指定的文件。
　　　在用 L 命令前,BX 和 CX 中应包含所读文件的字节数。

(10) 写命令 W,有两种格式：

　　1) 把数据写入磁盘的指定扇区：

　　　　－W　地址　驱动器　扇区号　扇区数

　　2) 把数据写入指定文件中：

　　　　－W ［地址］

　　　此命令把指定内存区域中的数据写入由 CS：5CH 处的 FCB 所规定的文件中。在用 W 命令前,BX 和 CX 中应包含要写入文件的字节数。

(11) 退出 DEBUG 命令 Q,该命令格式为

　　Q

　　它退出 DEBUG 程序,返回 DOS,但该命令本身并不把在内存中的文件存盘,如需存盘,应在执行 Q 命令前先执行写命令 W。

附录六 汇编程序出错信息

编码	说　　明
0	Block nesting error 嵌套过程、段、结构、宏指令、IRC、IRP 或 REPT 不是正确结束,如嵌套的外层已终止,而内层还是打开状态
1	Extra characters on line 当一行上已接受了定义指令说明的足够信息,而又出现多余的字符
2	Register already defined 汇编内部出现逻辑错误
3	Unknown symbol type 符号语句的类型字段中有些不能识别的东西
4	Redefinition of symbol 在第二遍扫视时,连续地定义一个符号
5	Symbol is multi-defined 重复定义一个符号
6	Phase error between passes 程序中有模棱两可的指令,以至于在汇编程序的两次扫视中,程序标号的位置在数值上改变了
7	Already had ELSE clause 在 ELSE 从句中试图再定义 ELSE 从句
8	Not in conditional block 在没有提供条件汇编指令的情况下,指定了 ENDIF 或 ELSE
9	Symbol not defined 符号没有定义
10	Syntax error 语句的语法与任何可识别的语法不匹配
11	Type illegal in context 指定的类型在长度上不可接收
12	Should have been group name 给出的组名不符合要求
13	Must be declared in pass 1 得到的不是汇编程序所要求的常数值,例如,向前引用的向量长度
14	Symbol type usage illegal PUBLIC 符号的使用不合法

续表

编码	说　　明
15	Symbol already different kind 企图定义与以前定义不同的符号
16	Symbol is reserved word 企图非法使用一个汇编程序的保留字
17	Forward reference is illegal 向前引用必须是在第一遍扫视中定义过的
18	Must be register 希望寄存器作为操作数，但用户提供的是符号而不是寄存器
19	Wrong type of register 指定的寄存器类型并不是指令或伪操作所要求的，例如 ASSUME AX
20	Must be segment or group　希望给出段或组，而不是其他
21	Symbol has no segment　想使用具有 SEG 的变量，而这个变量不能识别段
22	Must be symbol type 必须是 WORD,DW,QW,BYTE 或 TB，但接收的是其他内容
23	Already defined locally 试图定义一个符号作为 EXTERNAL，但这个符号已经在局部定义过了
24	Segment parameters are changed 对于 SEGMENT 的变量表与第一次使用该段的情况不一样
25	Not proper align/combine type SEGMENT 参数不正确
26	Reference to mult defined 指令引用的内容已是多次定义过的
27	Operand was expected 汇编程序需要的是操作数，但得到的却是其他内容
28	Operator was expected 汇编程序需要的是操作符，但得到的却是其他内容
29	Division by 0 or overflow 给出一个用零作除数的表达式
30	Shift count is negative 产生的移位表达式使移位计数值为负数
31	Operand type must be match 在自变量的长度和类型应该一致的情况下，汇编程序得到的并不一样，例如：交换
32	Illegal use of external 用非法的手段进行外部使用

续表

编码	说 明
33	Must be record field name 需要的是记录字段名,而得到的是其他东西
34	Must be record or field name 需要的是记录名或字段名,但得到的是其他东西
35	Operand must have size 需要的是操作数的长度,但得到的是其他内容
36	Must be var, label or constant 需要的是变量、标号或常数,但得到的是其他内容
37	Must be structure field name 需要的是结构字段名,但得到的是其他内容
38	Left operand must have segment 操作数的右边要求它的左边必须是某个段
39	One operand must be const 这是加法指令的非法使用
40	Operands must be same or 1 abs 这是减法指令的非法使用
41	Normal type operand expected 当需要变量标号时,得到的却是 STRUCT,FIFLDS,NAMES,BYTE,WORD 或 DW
42	Constant was expected 需要的是一个常量,得到的却是另外的内容
43	Operand must have segment SEG 伪操作使用不合法
44	Must be associated with data 有关项用的代码,而这里需要的是数据,例如用一个过程取代 DS
45	Must be associated with code 有关项用的是数据,而这里需要的是代码
46	Already have base register 试图重复基地址
47	Already have index register 试图重复变址地址
48	Must be index or base register 指令需要基址或变址寄存器,而指定的是其他寄存器
49	Illegal use of register 在指令中使用了 8088 没有的寄存器
50	Value is out of range 数值大于需要使用的,例如将 DW 传送到寄存器中

· 191 ·

续表

编码	说　　明
51	Operand not in IP Segment 由于操作数不在当前 IP 段中,因此不能存取
52	Improper operand type 使用的操作数不能产生操作码
53	Relative jump out of range 指定的转移超出了允许的范围(-128～+127 字节)
54	Index displ must be constant 试图使用脱离变址寄存器的变量偏移值
55	Illegal register value 指定的寄存器值不能放入"reg"字段中,(即"reg"字段大于 7)
56	No immediate mode 指定的立即方式或操作码都不能接收立即数,例如:PUSH
57	Illegal size for item 引用的项的长度是非法的,例如,双字的移位
58	Byte register is illegal 在上下文中,使用一个字节寄存器是非法的,例:PUSH AL
59	CS register illegal usage 试图非法使用 CS 寄存器,例如:XCHG CS,AX
60	Must be AX or AL 某些指令只能用 AX 或 AL,例如:IN 指令
61	Improper use of segment reg 段寄存器使用不合法。例如:立即数传送到段寄存器。
62	No or unreachable CS 试图转移到不可到达的标号
63	Operand combination illegal 在双操作数指令中,两个操作数的组合不合法
64	Near Jmp/Call to different CS 企图在不同的代码段内执行 NEAR 转移或调用
65	Label can't have seg override 非法使用段取代
66	Must have opcode after prefix 使用前缀指令之后,没有正确的操作码说明
67	Can't override ES segment 企图非法地在一条指令中取代 ES 寄存器,例如,存储字符串

续表

编码	说　　明
68	Can't reach with segment reg 没有做变量可达到的那种假设
69	Must be in segment block 企图在段外产生代码
70	Can't use EVEN on BYTE segment 被提出的是一个字节段,但试图使用 EVEN
71	Forward needs override 目前不使用这个信息
72	Illegal value for Dup count DUP 计数必须是常数,不能是 0 或负数
73	Symbol already external 企图在局部定义一个符号,但此符号已经是外部定义了
74	DUP is too large for linker DUP 嵌套太长,以至于从连接程序不能得到一个记录
75	Usage of ?（indeterminate）bad "?"使用不合适。例如,? +5
76	More values than defined with
77	Only initiallize list legal
78	Directive illegal in STRUC
79	Override with DUP is illegal
80	Field cannot be overridden
81	Override is of wrong type
82	Register can't be farward ref
83	Circular chain of EQU aliases
84	Feature not supported by the small Assembler（ASM）

附录七 IBM-PC ASCII 码字符表

低四位 B H \ 高四位 B H		0000 0	0001 1	0010 2	0011 3	0100 4	0101 5	0110 6	0111 7	1000 8	1001 9	1010 A	1011 B	1100 C	1101 D	1110 E	1111 F
0000	0	BLANK (NULL)	►	BLANK (SPACE)	0	@	P	`	p	Ç	É	á	▓	└	┴	α	≡
0001	1	☺	◄	!	1	A	Q	a	q	ü	æ	í	▒	┴	┬	β	±
0010	2	☻	↕	"	2	B	R	b	r	é	Æ	ó	▓	┬	├	Γ	≥
0011	3	♥	‼	#	3	C	S	c	s	â	ô	ú	│	├	─	π	≤
0100	4	♦	¶	$	4	D	T	d	t	ä	ö	ñ	┤	─	┼	Σ	∫
0101	5	♣	§	%	5	E	U	e	u	à	ò	Ñ	╡	┼	╞	σ	⌡
0110	6	♠	▬	&	6	F	V	f	v	å	û	ª	╢	╞	╟	μ	÷
0111	7	•	↨	'	7	G	W	g	w	ç	ù	º	╖	╟	╠	τ	≈
1000	8	◘	↑	(	8	H	X	h	x	ê	ÿ	¿	╕	╚	╔	Φ	°
1001	9	○	↓	)	9	I	Y	i	y	ë	Ö	⌐	╣	╔	╦	Θ	•
1010	A	◙	→	*	:	J	Z	j	z	è	Ü	¬	║	╩	╦	Ω	·
1011	B	♂	←	+	;	K	[	k	{	ï	¢	½	╗	╦	■	δ	√
1100	C	♀	∟	,	<	L	\	l			î	£	¼	╝	╠	■	∞	ⁿ
1101	D	♪	↔	-	=	M	]	m	}	ì	¥	¡	╜	═	■	φ	²
1110	E	♫	▲	.	>	N	^	n	~	Ä	₧	«	╛	╬	■	∈	■
1111	F	☼	▼	/	?	O	_	o	△	Å	ƒ	»	┐	╧	■	∩	BLANK FF